Resource Recovery from Municipal Solid Wastes

Volume I
Primary Processing

Authors

Luis F. Diaz
President
Cal Recovery Systems, Inc.
Richmond, California

George M. Savage
Vice President
Cal Recovery Systems, Inc.
Richmond, California

Clarence G. Golueke
Director of Research and Development
Cal Recovery Systems, Inc.
Richmond, California

CRC Press, Inc.
Boca Raton, Florida

Library of Congress Cataloging in Publication Data
Main entry under title:

Resource recovery from municipal solid wastes.

Bibliography: p.
Includes index,
Contents: v. 1. Primary processing—v. 2. Final processing.
1. Recycling (waste, etc.) 2. Refuse and refuse disposal. I. Diaz, Luis F. II. Savage, George, 1913- . III. Golueke, Clarence G., 1917-
TD794.5.R455 628.4′45 81-18021
ISBN 0-8493-5613-X (v. 1) AACR2
ISBN 0-8493-5614-8 (v. 2)

Direct all inquiries to CRC Press, 2000 Corporate Blvd. N.W., Boca Raton, Florida, 33431.

International Standard Book Number 0-8493-5613-X (Volume I)
International Standard Book Number 0-8493-5614-8 (Volume II)

Library of Congress Card Number 81-18021
Printed in the United States

THE AUTHORS

Dr. Luis F. Diaz received his B.S. degree in Mechanical Engineering from San Jose State University. He then entered the University of California (Berkeley) where he received his M.S. and Ph.D. degrees in the area of Environmental Engineering. Dr. Diaz was instrumental in developing the solid waste processing facility at the Richmond Field Station of U.C. Berkeley. During the course of developing this facility, Dr. Diaz played a major role in developing the fiber recovery system for extracting cellulose fiber from municipal solid waste. Dr. Diaz then undertook an extensive study of biogasification of various waste fractions. This work dealt with the digestion of organic waste fractions and sludges. This work later expanded to include the study of integrated energy-agro-waste systems. This dealt primarily with the use of waste heat as an energy source in an agricultural complex integrated with a wastewater treatment facility.

Dr. Diaz now serves in the capacity of President of Cal Recovery Systems. He has served as a consultant to various municipalities and governmental agencies, as well as leading the work of Cal Recovery Systems in various international projects. He has given numerous invited lectures at technical meetings and has served as the co-organizer of several seminars. He has also worked as consultant to international agencies such as the United Nations (WHO and UNIDO), the World Bank, the Asian Development Bank, and the U.S. Agency for International Development.

George Savage was born in 1948 in Faribault, Minnesota. He received his B.S. and M.S. degree in Mechanical Engineering at the University of California (Berkeley). He currently serves as Vice President of Cal Recovery Systems. He has been actively involved in the development of waste processing technology. He conducted pioneering efforts in the areas of developing test methods and procedures and of testing of waste communition, air classification, screening, and densification systems. He has conducted field testing programs at all of the major resource recovery plants in the U.S. In addition, he has advised equipment manufacturing companies on improvements and devices for recovering various waste fractions from the solid waste stream.

He is a member of several technical societies and has served as a lecturer on various aspects of solid waste management at several national conferences.

Dr. Clarence G. Golueke, Ph.D., has been the Director of Research and Development at Cal Recovery Systems, since 1978. Prior to that date he was a member of the faculty of the Division of Sanitary Engineering of the Department of Civil Engineering, University of California.

Dr. Golueke received his B.A. degree from St. Louis University, Illinois and his M.A. degree from the University of Illinois in Urbana. He received his doctorate of philosophy from the University of California at Berkeley.

He is a member of several professional and technical societies and the honor society of Sigma Xi. He was the recipient of the A.M. Wellington Award (American Society of Civil Engineers) in 1966.

His professional activities include service as consultant to federal, state, and local governments and to industry on various aspects of solid and liquid waste management and environmental control.

Dr. Golueke is the author or co-author of more than 100 publications, a large portion which are concerned with biological systems of resource recovery from solid and liquid wastes. He is also on the boards of five scientific journals.

RESOURCE RECOVERY FROM MUNICIPAL SOLID WASTES

Luis F. Diaz, George M. Savage, and Clarence G. Golueke

Volume I
Primary Processing

Volume II
Final Processing

TABLE OF CONTENTS

Chapter 1

INTRODUCTION

I. INTRODUCTION

As recently as 1980, it would have been exceedingly difficult to have found a point about which the public consensus was as great as that existing on the desirability and even the necessity for resource recovery. Even in the politically conservative climate of 1981, the desirability continues to be almost unanimously recognized, although feelings regarding the degree of urgency vary widely. The variation is manifested by the diversity of opinions regarding the intensity of effort and the extent of concessions and even sacrifices to be made in carrying on resource recovery. Certainly, the more urgent one regards an activity as being, the greater is the effort willingly expended and sacrifices made to ensure its successful outcome. On a public scale, urgency translates itself into priorities in undertakings. Logically, the more urgent the need, the higher the priority accorded it. Priority may be expressed in the form of extent of practice and through special consideration accorded the practice in question. Judged on the basis of the first criterion, as of 1978 recycling had a low priority, inasmuch as in that year overall recovery amounted to less than 7% of the gross municipal discards, and that mostly in the form of reclamation of paper.[1] Examples pertaining to the second criterion are the allowing of exceptions, the granting of a subsidy, and the imposition of a legal specification. A very practical manifestation of the existing wide range of estimates of urgency may be found in the long-standing dispute between the National Association of Secondary Materials Industries (NASMI) and the Interstate Commerce Commission (ICC) regarding freight rates. The priority given resource recovery by the ICC apparently is so low as of this writing that the agency would like to permit higher rates to be charged for transporting secondary materials than for virgin materials. On the other hand, the NASMI and resource conservationists continue to press for at least an equalization of rates, and hopefully even preferential treatment.

II. FACTORS AND PROBLEMS IN RESOURCE RECOVERY

Lest one be unduly critical of past and present inaction, several factors, or more aptly termed "problems," exist which combine to impede progress in resource recovery. One of the problems, and a very serious one indeed, is the disparity between the requirements regarding the quality of secondary materials intended for energy production and for use in the manufacture of new items, and the nature and quality of these materials as they exist in the municipal waste stream. The requirement of homogeneity of feed material in energy production and in the manufacture of useful products as contrasted to the heterogeneity of municipal wastes exemplifies this disparity.

Homogeneity, as it applies to wastes destined for use in thermal energy recovery, implies that the wastes be combustible and free of substances that are not combustible, or can interfere with combustion, or can corrode the energy conversion unit either directly or indirectly. As applied to energy recovery by biological means, the term implies the absence of biologically nondegradable materials and of substances that are toxic to microorganisms. The term has a somewhat narrower connotation when applied to the recovery of materials from wastes for use in the maufacture of useful products. To be used in the manufacture of a given product, a substance must be in the form of a uniform mass, that is, it must be characterized by a certain degree of purity. Taken in its broad sense, homogeneity implies that all items in a given mass are comparable in composition, e.g., aluminum beverage containers, all-steel containers, etc. Moreover, the composition of the materials used in manufacturing the items should be identical in all items. Thus, the constituents of alloys should be the same, or at least should be known. Taken in the narrow sense, homogeneity refers to single elements (e.g., ferrous metal, paper fiber).

As far as the manufacturer and the energy producer are concerned, the difficulty with establishing an industry based upon the use of secondary materials scavenged from municipal waste is not confined to the shortcomings described in the two preceding paragraphs. Four additional factors can be adduced, each of which has the potential of exercising a decisive role in determining the success or failure of an undertaking. Briefly stated they are (1) assurance of long-term and uninterrupted availability of the material to be recovered; (2) availability of the material at a price competitive on an overall basis with virgin (primary) material; (3) distances to and between sources of the secondary material are such that transportation costs are not excessive; and (4) transportation (freight) rates must at the least not be higher than those for virgin materials.

A. Need for Homogeneity

The need for homogeneity in thermal energy recovery ultimately relates to the requirement that the recovered waste material be economically competitive with fossil energy sources. To be so, the waste material must have a reasonably high heating value and must be adaptable for use in existing energy conversion systems. The required characteristics are explained in some detail in the sections on energy recovery. The need for homogeneity, as it relates to biological energy recovery, rests upon the fact that the waste material must serve as the principal substrate in the culture of the microorganisms involved in the energy recovery process. The ramifications of this function are elaborated upon in the section on biological energy recovery. Inasmuch as the energy implications of homogeneity are amplified in subsequent sections, the remainder of this section is devoted to homogeneity as it relates to waste materials destined for use in the manufacture of products.

The origin of the need for homogeneity in the manufacture of products is not hard to determine. Uniformity of quality and specifications of a product in a manufacturing operation demand precise control over the raw materials used in the process. Obviously, one of the essential factors in precise control is a knowledge of the identity of the components of the raw material, as well as the capability of feeding those components into a process at carefully regulated rates and amounts. The further the deviation from this ideal, the lower will be the quality of the manufactured product. The quality of certain types of products, especially those made of plastic or of aluminum, are especially sensitive to degree of homogeneity. In fact, with plastic, the very act of refashioning the material in a recycling operation results in a product having a quality lower than that of the original.

The need for cleanliness stems from the requisites, homogeneity and purity, inasmuch as uncleanliness bespeaks contamination with foreign substances. In certain manufacturing processes, the presence of even minute amounts of foreign materials drastically and adversely affects the quality of the product. For example, trace amounts of copper in iron scrap reduces the tensile strength of the steel. In the manufacture of paper from reclaimed fiber, plastic, or dirt particle contaminants can disrupt the paper manufacture process and seriously detract from the utility of reclaimed paper as a raw material. For example, according to the U.S. Environmental Protection Agency, (EPA), the manufacture of paper products from paper that had been in contact with pathogenic material may result in a new product not hygienically safe for use in packaging food.

In the preceding paragraphs a case was built up for the need of a raw material that is homogeneous in nature and composition and which exists in a clean state. In this paragraph is described the spread between the quality of the material in the waste stream and that which would be ideal. A collection of tables on the composition of municipal refuse is not needed to demonstrate the great gap between the ideal and the actual, because a visit to a local transfer station or dump would overwhelmingly reveal the enormity of the disparity. Among the many types of wastes generated by human activity, municipal waste is the most heterogeneous in composition. The heterogeneity is inevitable since municipal waste is a conglomeration of rejects from everything used in typical urban existence. Not only is the waste a heterogeneous mass, but the components are intermingled in a seemingly hopelessly intertwined tangle. As if the heterogeneity were not sufficient to discourage the prospective scavenger, the sheer dirtiness of the mass is dismaying. Joined with the heterogeneity and filthiness of the waste is a third important characteristic, namely abrasiveness. Because of the unusually severe abrasiveness of municipal waste, wear and tear on equipment used in processing it is greatly intensified. The combination of the preceding three characteristics has heretofore practically ruled out centralized resource recovery, excepting perhaps for the magnetic removal of ferrous products.

At this point in the discussion, it might be said that source separation could eliminate or minimize the problems discussed in the preceding paragraphs, and perhaps rightly so, as is shown in a later section. The fact remains, however, that thus far the practice of source separation has been almost infinitesimal in extent. Partial explanations for rarity of source separation in the past and present are inconvenience for the householder and alleged reduction in efficiency of collection and a consequent increase in cost. Nevertheless, the principal reason is the fact that to the public-at-large and hence to politicians, the need to recycle seems to be less than imperative. This latter reason, in turn, takes away from the average citizen a prime motive for the additional exertion required to classify and store his or her daily output in a collection of separate containers. Of course, factors other than the preceding will come into play as resource recovery becomes more widely practiced.

B. Long-Term Uninterrupted Availability of Items

This requirement is an obvious one and rests upon the fact that it would be folly to build an industry on an uncertain supply of raw material. With the existing rate of municipal waste generation, the response as to the fulfillment of this requirement would almost unanimously and immediately be one of positive and unequivocal assurance. The fact is that as long as our present way of life persists, waste will continue to be generated at least at the present rate. The question, however, is not one of total amount, but rather of the individual components that together make up the total mass. An example is the steel container, i.e., the "tin can". Here, gain in the fraction of food products marketed as frozen foods is reflected by a decline in the amount marketed in steel cans, and consequently in a drop in number of cans discarded. Additionally, if the present move towards the compulsory use of returnable beverage containers takes hold, the usage and subsequent discard of steel cans will drop. The fact that at present the use of steel cans in the beverage industry is far less than that of aluminum cans does not alter the overall situation.

Related to availability of given waste materials is the question of ownership of the materials. Despite the fact that for the present almost every municipality is eager to give a part or even all of its waste stream to any and all takers, provided of course the taking is done under proper and controlled conditions, the continued duration of that magnanimity is not necessarily assured. Chances are that if the industry presently receiving the material *gratis* should prosper, the citizenry upon becoming aware of the fact would begin to question the community's apparent munificence to the industry. Undoubtedly, the questioning would be followed by the imposition of a charge for the material, and the financial benefit accruing from the use of the latter would be correspondingly less. In summary then, the ownership of the waste resources and the future policies of the owners with respect to the wastes must be unequivocally established before a candidate industry is willing to commit itself to the utilization of the waste resource.

Unfortunately, because the nature and amount of materials discarded are influenced by the vagaries of man's activities and the fluctuations in his economic well-being, extrapolations as to the extent of availability of given components must be attended by a high degree of uncertainty. The same uncertainty applies to predictions concerning the future attitude of the public regarding access of a resource recovery industry to a waste that has developed a market value because of the economic success of an enterprise based upon its (the waste) utilization. Obviously, the processor's interest in the components will come to an abrupt end when he finds that the cost of the primary resource is lower than that of the secondary material.

C. Reasonable Transportation Cost

The requisite, reasonable transportation costs, pertains to degree of concentration of the sources of the waste and of accessibility to the sources. Expressed in a negative way, the sources of the component wastes must not be so scattered or so distant that the costs of transporting them to the user significantly exceed the cost of transporting primary material. This requisite probably is the most difficult one to meet in resource recovery. As would be expected, the largest individual concentrations of wastes, and hence of given waste components, are to be found near areas in which the human population is at its densest. While these latter areas may contain a sizeable percentage of the nation's population, the larger fraction is to be found in small to medium-sized communities across the country. Consequently, a large portion of the nation's discarded resources also is widely scattered. Unless an economically viable means of collecting the latter wastes is found, a significant portion of recoverable resources will have to be by-passed.

The problem of scattered waste disposal sites can be alleviated considerably by establishing a regional approach to solid waste management. This can be done through statewide planning in which is provided a statewide policy for resource recovery, much as was done by the state of Wisconsin.[2] Of course, even with the establishment of regional centers, the need would remain for developing an effective transportation system. Probably, a combination of rail and truck haul would be the best approach.

The generally appreciable distance between the location of the potential user of a reclaimed resource and of that where municipal wastes are generated and disposed, is the source of yet another difficulty as far as materials recovery is concerned. As one would expect, the most efficient and hence cost-effective approach in the manufacture of a product is to locate the manufacturing plant as closely as possible to the source of its raw material. This is precisely what happens in the paper industry, in that paper manufacturing plants generally are located in close proximity to the forests from which the pulpwood used in paper manufacture is obtained. If more than one raw material is used in an industry, then a compromise location is selected, but even here the site of choice almost invariably is not in close proximity to large population centers. Consequently, it is not surprising that an acute problem arises when an attempt is made to substitute secondary materials for all or even only a part of the primary materials hitherto used in an industry. In most instances the secondary material must be transported over a longer distance to reach the plant than is the case with primary material. The latter problem becomes especially severe when a switch is made from primary to secondary fiber in an existing paper manufacturing plant. However, the transportation problem will disappear when older plants become obsolete and are replaced by modern versions. The reason is that part of the renovation would include relocation to a site more accessible to centers of waste generation.

D. Ability of Recovery Process to Accommodate Secondary Materials

Transporting the secondary material to the manufacturing plant or energy production facility is only a part of the difficulty. Once at the plant or facility, the latter must have the capacity to accommodate secondary as well as primary materials (e.g., fossil fuels). Accommodating secondary materials almost invariably involves modification of existing processes and, hence, of equipment. The extent of the required modification may be quite great. Fortunately, most existing manufacturing and energy conversion processes are designed such that they can incorporate at least some secondary materials. Thus, in thermal energy conversion, processed municipal wastes can be used as a substitute for coal. In glass making the use of cullet is an essential part of the process. In paper making, rejected paper is recycled by way of the pulpers, and in steel making, scrap in the form of rejects is used along with iron ore. However, it should be pointed out that in the three immediately preceding examples, manufacturing rejects generally are used. In terms of homogeneity and cleanliness, manufacturing rejects are a far cry from rejects (wastes) reclaimed from the municipal waste stream.

A handicap of governmental origin is constituted by the differences between the treatment accorded the use of secondary material and that given to the use of primary materials. On the one hand, the Federal government subsidizes the iron ore industry, actually a subsidiary of the steelmaking industry, through tax deductions in the form of extremely generous depletion allowances. On the other hand, no such subsidization is accorded the secondary materials industry. The resulting handicap in terms of disparity of costs is substantial.

E. Government Assistance

While the nature of the course of action required to rectify inequities may be straightforward and quite obvious, that of providing governmental assistance is not as clear-cut. One approach could consist in the provision of some type of a subsidy. A relatively straightforward subsidy is the tax break. The first step, of course, in allowing tax breaks would be to establish a program that would result in a monetary assistance to compensate for any increase in cost that may attend the use of secondary materials. Moreover, the long-term benefits resulting from the use of secondary materials, namely, conservation of resources and lightening of the solid waste burden, justify the assistance required to expand the use of secondary materials, even to the extent of allowing a reasonable economic advantage over the use of primary materials.

A second form of governmental assistance could be a price support system. Arguments for and against price support are quite well summarized by G. S. Gill in an article in *Compost Science*.[3] According to Gill, an important feature of a price-support program is its apparent flexibility. The support "can be instituted, expanded, contracted, or even withdrawn without causing too much ado."[3] Of course, the costs involved would not only be those of the price-support plan itself, but also those of administering the plan. The experience with agriculture price-support systems would be useful in determining the magnitude of the administrative costs. Gill probably is somewhat overly optimistic in his statement, "Even though institutional rigidities will develop in time, even in relation to as flexible a program as the price-support programs—these obstacles will be relatively easy to overcome inasmuch as there are no sunk costs involved."[3] The stiff resistance put up by interested parties in the past few years to governmental attempts to drastically reduce price-support for certain agricultural commodities tends to belie Gill's predictions as to ease of withdrawing price supports. Nevertheless, as he states, price-support programs do constitute an incentive program that would merit investigation.

The solution to governmentally oriented problems seems obvious, namely positive action at least to remove existing inequities. The removal would seem to be a rather straightforward task, but the unsuccessful outcomes of attempts by the secondary materials industry to bring about a redress of their complaints has proved to be an arduous undertaking attended by only a modicum of success. The reason is the intense opposition of powerful industries committed to the use of primary materials.

Another problem area is that of marketing. Ultimately the source of the difficulty is the nature of the municipal waste from which the secondary materials are extracted. The effects of the nature of municipal waste were discussed in the preceding paragraphs under the guise of difficulties associated with the nature of the reclaimed materials and the attendant factors (e.g., "dirtiness", nonhomogeneity, uncertainty of continued supply) and their bearing on the utilization as raw material in a particular industry. These difficulties combine to make the reclaimed items less attractive to potential consumers.

F. Public Attitude

A factor not covered in the preceding paragraphs and which is exceedingly influential in marketing is public attitude. With the average individual, the prospect of using a "secondhand" item brings with it a feeling of repugnance or at least, of reluctance. This feeling is the result of a cultural heritage in which the use of salvaged goods is associated with economic distress, unless the item happens to fit under the classification "heirloom" or "antique". The feeling also stems in part from a conviction, often based on fact, that reclaimed items, because of having already been subjected to use, have thereby lost some of their original utility or durability. However, this feeling is more apt to occur when the discarded item is used directly (e.g., "second-hand" clothing), than when it is processed to become a raw material for the production of a different item. This cultural problem area in marketing will diminish as potential consumers become aware of the fact that there is a considerable difference between using a discarded item unaltered (except for a certain amount of refurbishing) and processing it to serve as a raw material in the production of a new item. The difference is more pronounced with the reuse of metals than with fibers. The latter do deteriorate significantly with each reuse. Unfortunately, while all that has been stated in these sentences may be true and most individuals are convinced of the facts, translating the conviction into action has a long way to go.

The public attitude is much more positive towards the use of wastes in energy recovery. The favorable attitude undoubtedly stems from the public's awareness of the critical energy situation. Not to be overlooked is the fact that in energy recovery, the contact between the user of the recovered energy and the raw waste is far removed.

G. Technological Weaknesses

The final obstacle to progress in the reclamation of discarded resources owes its origin to the relative lateness of the concern about conservation and the consequent interest in recovering discarded resources. Because of the lateness, existing technology with which to accomplish resource recovery still leaves much to be desired. To remedy the situation, numerous schemes and types of equipment have been developed and proposed. Interestingly, a sizeable number are based on concepts and equipment designs that were in existence years ago, although never on an appreciable scale. For example, the Lantz Converter, a pyrolysis unit, was being promoted 20 years ago. The concept of using waste heat from an incinerator to generate steam is described in a textbook published in 1901.[4] The technology of heat recovery presently in vogue, although incorporating sophisticated design improvements, nevertheless has yet to overcome many of the failings of the earlier version described in the 1901 textbook.

The rapid expansion of the technology and the feeling of urgency to do something about conservation have combined to provide a climate in which is fostered a naiveté on the part of decision makers, professionals, and the citizenry in their respective evaluations of the many processes and equipment designs currently being touted as the answer to existing needs. This naiveté coupled with the strong promotional tactics of enthusiasts and entrepreneurs have led to rash undertakings that later turned into expensive failures. Damage by way of an ill-conceived venture is not only economic in nature but is also is a loss of credibility and a severe dampening of enthusiasm.

Unfortunately, it is extremely difficult to find informed, critical evaluations of the several systems currently in vogue. To be of greatest use, an evaluation should include: (1) an in-depth, objective analysis of the potential of the system in question and of its advantages and disadvantages; (2) a searching discussion as to the applications to which it is or is not suited; and (3) a critical comparison with alternative systems, e.g., direct burning (refuse-derived fuel) vs. pyrolysis.

III. OBJECTIVES AND CONTENTS OF THE VOLUME

It was with the hope of meeting the need for up-to-date critical evaluations of currently publicized recovery systems, as well as to suggest approaches to alleviating the several problems described herein, that this and the succeeding volume were compiled. It would be presumptuous on the part of its authors to expect that the two publications would fully meet all existing needs, but the conviction is justifiable that they will provide a groundwork whereby the reader can go about the task of arriving at a reasonable conclusion.

To facilitate the presentation of so broad a subject, and in keeping with the format recommended by their publisher, the authors decided to divide their presentation into two volumes. Accordingly, Volume I deals with the collection and preparation of the raw material (municipal solid wastes); and Volume II, with the reclamation and, if necessary, the disposal of the collected and prepared wastes. Thus, Volume I begins with a brief overview of the three steps conventionally involved in the transfer of municipal waste from its generators to the processing facility or disposal site, namely, storage, collection, and transport. (Transport constitutes a separate step when the use of a transfer station is required.) Thereafter are described and discussed in the remainder of Volume I and continued in Volume II, the principles and steps involved in the preparatory and final planning, designing, operation, and environment implications of a resource recovery facility as a whole. In the course of the presentation, the terms "front-end systems" and "back-end systems" are defined and explained. Thus, "front-end systems" are defined as the collection of unit processes through which incoming raw wastes are sorted and processed to become an assortment of feedstocks for matching "back-end systems". The definition of "front-end systems" leads to that of "back-end systems" in that the latter are systems designed and operated to convert the feedstocks from front-end systems into a useful product, whether it be energy or a potential raw material for a manufacturing process.

In addition to the storage, collection and transport named in the preceding paragraph, Volume I deals with the unit processes that make up front-end systems. In it are described process and design considerations pertinent to shredding, air classification, and screening. The presentation on the back-end systems selected for discussion is made in Volume II. In the volume are described representative biological and nonbiological systems for energy and materials recovery. Selected biological systems related to energy recovery are anaerobic digestion to produce methane, cellulose hydrolysis-ethanol fermentation, and collection of biogas from landfills. Composting and single-cell protein production are the two biological materials recovery systems. Incineration and refuse-derived fuel (RDF) production are the two nonbiological systems selected for description. Nonbiological systems for the recovery of materials that receive attention are those for paper, glass, aluminum, and magnetic (ferrous) metals. Volume II includes a chapter on the environmental aspects of resource recovery activities. By including the chapter, the authors hope to instill the reader with the feeling that preservation and improvement of the quality of the environment should be key considerations in the evaluation of each and every resource recovery program. Discussion of the sanitary landfill appropriately is reserved for the final part of Volume II, that is, after resource recovery has received its due attention. It is almost a truism that no matter how efficient and effective a resource recovery program may be, a waste or residue will remain for which the only practical destination is the landfill. Aside from that extreme, political, economic, and technological conditions may individually or collectively render impractical the recovery of certain resources discarded as waste materials.

The mention of trade names or commercial products does not constitute endorsement or recommendation for use.

REFERENCES

1. U.S. EPA, Fourth Report to Congress: Resource Recovery and Waste Reduction (SW-600), Office of Solid Waste, U.S. EPA, Aug. 1, 1977, 142.
2. **McGauhey, P. E., Koerper, E. C., and Wisely, F. E.,** Wisconsin Solid Waste Recycling—Predesign Report, report prepared by a Board of Engineering Consultants for the Governor's Recycling Task Force, State of Wisconsin, May 31, 1973.
3. **Gill, G. S.,** Public Price Support as an Incentive to the Use of Solid Waste as a Resource, *Compost Sci.,* 12(1), 16, 1971.
4. **Goodrich, W. F.,** *The Economic Disposal of Town's Refuse*, P. S. King and Son, London, 1901.

Chapter 2

STORAGE, COLLECTION, AND TRANSPORT

I. INTRODUCTION

The implementation of a decision to institute a program of resource recovery involves more than the mere addition of one more step to the series of the many that constitute the management of municipal solid wastes. On the contrary, the ease and effectiveness of the implementation is affected by each of the steps in the series. This fact is especially so in the storage, collection, and transport of municipal solid wastes. The three steps can either simplify or complicate resource recovery. In general practice, the result usually is one of complication in that recyclable resources and nonrecyclable wastes are indiscriminately intermixed. The mixing begins with storage and is compounded in collection and transport.

Although resource recovery is the principal concern of this book, a brief overview is presented of conventional storage, collection, and transport, mostly because the steps do have an important bearing on resource recovery. A lesser reason is to provide an opportunity for evaluating so-called advances made in the technological aspects of the three steps. A more complete treatment of the subject can be found in a solid waste textbook.[1-3]

II. STORAGE AT POINT OF ORIGIN

The need to store municipal solid waste at the point of origin arises from the discontinuous nature of solid waste collection. If a waste cannot be removed continuously, obviously, it must be stored until it can be transported to the site of disposal or processing. Under conditions characteristic of present communities, pipe transport is the only feasible means of removing a waste on a continuous basis. But pipe transport places a premium upon homogeneity, imposes an upper limit on particle size and density, and demands some degree of fluidity. On the other hand, an outstanding characteristic of municipal solid waste is its heterogeneity, its having a sizeable fraction with one or more dimensions approaching a meter or more, and a practically complete lack of fluidity. Consequently, unless subjected to a considerable processing, municipal solid waste does not readily lend itself to pipe transport, and hence, to continuous removal.

A. Principles

A fundamental principle in attaining satisfactory storage is to isolate the waste from the environment. It follows then that a prime measure of satisfactory storage is degree of isolation from the environment. The required extent of isolation is a function of the nature of the waste, that is, of the degree to which the waste, if exposed, could adversely affect public health or the quality of the environment. Thus, the amount of isolation to be provided in the storage of a highly putrescible waste would be far greater than that for a relatively stable material such as paper or metallic objects.

Additional requirements include that stored wastes should occupy a minimum of space; space occupied by wastes is nonproductive, and in a commercial establishment it represents a loss of income. The waste should be readily accessible for collection; this requirement ensures an efficient employment of manpower and tends to minimize spillage. The costs involved should be the lowest commensurately with hygiene and acceptable aesthetics.

B. Containers

Wastes are stored in suitable containers. A general requirement for a container is that it be easily handled and that it can be closed. Specifically, a container for domestic refuse should be fabricated of a durable material (metal or rigid plastic) and should be equipped with a tight-fitting lid. Its capacity when filled should be such that it can be safely manipulated by the collection crew. Generally, with combined collection (rubbish and garbage in one container), a capacity of about 115 ℓ is a safe one. If the garbage fraction is segregated and collected separately, the maximum volume of its container should be 45 ℓ.

Within the past decade or so a trend has been established in which a plastic-lined paper bag or a bag solely of plastic is substituted for the metal or rigid plastic container. The bag, which should have a capacity roughly equivalent to that of a 115-ℓ rigid container, is suspended until filled on an especially designed rack on the householder's premises. When filled, the bag is suitably closed and is ready for collection. The advantages of bag storage are predominantly in terms of ease of collection. For example, with bag storage, the total weight of container (exclusive of contents) lifted by a crew member on a typical day would be about 60 kg, whereas with conventional containers, the total would amount to 600 kg. Collection time is shortened because less handling is involved in that the bag simply is tossed into the collection vehicle, whereas a rigid container must be emptied and then returned to its former position.

Of course, the use of bag containers is not without its serious disadvantages. For instance, a bag is not as durable as a rigid container, and the cost to the householder is increased. However, the most serious disadvantage develops at the processing site. In a resource recovery operation, the bag must be opened, ruptured, or split before its contents can be processed. The bag itself, being of plastic film (or if paper, contains a plastic film liner), is size reduced with difficulty, and in general constitutes a handling problem.

Other developments in container design are in the nature of adaptations to collection vehicles that depart from conventional design.

Commercial wastes are best stored in a large metal container equipped with a hinged lid. The container is designed such that it and its contents can be transported to the disposal site. Containers are available for use in conjunction with compaction equipment, thereby making it possible to increase the capacity of the container as much as fourfold.

A relatively recent development is the use of containers as "miniature transfer stations" in rural areas. The containers are strategically placed such that they are convenient to the households they are to serve, and yet can be readily transported to the disposal site. Problems encountered with such an arrangement generally result from insufficient cooperation on the part of the people benefiting from the service.

This section on storage should not be concluded without calling attention to the importance of arranging for adequate and easily accessible storage facilities in the design of commercial buildings. The tendency is for the architect and his or her client to overlook this aspect, probably because of a reluctance to dedicate valuable space to the storage of wastes. The consequences of failing to make a proper provision are felt throughout the useful life of a building, and more than outweigh the illusory benefits from not doing so.

III. COLLECTION

This section is concerned almost solely with the collection of residential refuse. The collection of domestic refuse is a more visible operation and is more prone to problems than is the case with commercial wastes. The key problems in the collection of residential refuse arise from the three factors—labor, traffic, and costliness.

By its very nature, collection is labor intensive. Not only is it subject to all of the problems that attend any labor-intensive operation, it has certain very serious ones peculiar to it. These latter all stem from the fact that the collector's task is strenuous, poses certain hazards, and leads to an increased incidence of arthritis and a susceptibility to coronary heart disease. Moreover, the job certainly does lack aesthetic appeal, and is popularly regarded as being somewhat déclassé. The strenuous nature of the job is indicated by the following account. It has been estimated that during a typical day, each member of a collection crew will have walked from 16 to 19 km and will have hauled slightly more than 2 Mg of material to the collection vehicle. That refuse collection is characterized by a certain amount of hazard, is attested by its high injury frequency rate. For example, in a study of the health and safety record of refuse collection in the city of New York,[4] it was observed that frequency of injury rate (disabling injuries per one million man-hr) was 20 times that in industry in general and 5 times that in underground mining. The relative positions of severity rates (days lost per one million man-hr) were somewhat different in that the severity rate in sanitation was about twice that in industry in general, but about one fourth that in underground mining. The increased incidence of disease is exemplified by the fact that the coronary heart disease rate among sanitation workers is almost double that among the population in general.[4] The number of sanitation workers afflicted with disabling arthritis is about 1.33 times that of farmers, and from four to five times that of the populace at large. The factors of aesthetics and low social status at least until recently are self-evident.

The hazard problem can be lessened both by improvements in the design of the collection vehicle and by adequate training of the collection crew. The increased incidence in coronary heart disease and arthritis is difficult to change because of the nature of the collection task. Improvement in container design and the use of packers have reduced the objectionable visual and olfactory impact. The déclassé can be changed for the better by convincing the public of the importance of refuse collection, and perhaps by providing clean uniforms for the workers. Finally, a suitable pay scale is the best compensation for an occupation as generally unattractive as is refuse collection. Attempts made to reduce the dependence on labor have ranged from limiting the crew size to a single individual to the incorporation of automation features into the design of the collection vehicle. These features are described in the section on vehicles.

In the U.S., the crew generally consists of two or three individuals, one of whom operates the vehicle. Although in the past this arrangement was quite satisfactory, for the reasons cited in the preceding paragraph several communities are trying the one-man crew approach, apparently with some success. However, the trend is as yet far from widespread. To make the transition to a one-man crew, certain adjustments must be made. Among them are the use of bags for storing refuse, curbside collection, and of collection vehicles equipped with the steering wheel and the loading hopper on the right-hand (curb) side of the vehicle. Weight limitations on containers must be strictly enforced. Moreover, to get the required cooperation, the wages of the single crew member must be higher than that paid the individual members of a three-man crew. Interestingly, the time requirement in man-min/Mg of refuse loaded remains almost the same regardless of whether the crew consists of one or two men.[5]

Problems associated with traffic occur because collection is carried on with the use of surface vehicles that must traverse public thoroughfares. Even by adjusting collection schedules to off-peak traffic hours, competition with other vehicles prolongs travel time. Time spent on the road extends crew time and ties up an expensive piece of equipment, namely, the collection truck. Other effects are an aggravation of the overall air pollution problem and the imposition of yet another drain upon the fossil fuel supply. A further, albeit indirect, effect is a contribution to the many difficulties arising from other sources in the siting of a processing facility. The difficulty comes from the objections of affected citizens to the converging traffic of the collection vehicles at the processing site.

Collection is an expensive operation and is becoming more so because it is especially affected by the inflationary spiral. A decade ago (i.e., 1970) the cost per metric ton of refuse collected was on the order of $18 to $25, which translates into $45 to $120 per household.

A. Considerations

Planning a collection system entails a number of considerations, among which are combined vs. separate collection, frequency of collection, point of collection, crew size, pickup density, programming, and equipment. Since reference already had been made to crew size, in this section it is discussed only incidentally as a part of the other features.

1. Separate Vs. Combined Collection

Separate collection in the "classical" sense simply means a segregation of garbage (food preparation wastes) from rubbish both during storage and in collection. The popularity of the separate mode of collection disappeared when circumstances compelled the cessation of feeding garbage to swine. A contributing factor was the resistance of the population-at-large to separate collection. Separation in the classical sense facilitates resource recovery to the extent of reducing the contamination of recyclable materials with putrescible organic matter. Separation would certainly be advantageous in a program involving biological reclamation, because the readily decomposable organic matter would be relatively free of refractory organic materials and of inorganic (inert) components. A good example of such a biological reclamation would be composting, in that the segregated food wastes would have the requisite carbon-nitrogen ratio.

An extension of separate collection has become known as "source separation". It receives attention in another section.

2. Frequency

Among the several factors that determine frequency of collection are type of collection (i.e., combined vs. separate), volume of generation, composition (nature) of the waste, effect on rate of waste generation, cost, and fly production. Type of collection influences frequency in that if the garbage fraction is stored separately from the rubbish, the garbage should be collected each day because it is very putrescible and serves as a haven for flies in all their life cycle forms. On the other hand, the rubbish, being uncontaminated by garbage and being quite stable, can be stored for a relatively long time without giving rise to nuisances or public health problems. Consequently, the frequency of rubbish collection is a function mainly of rate of generation and costs. Volume of generation is a self-evident factor in that the indicated frequency of collection is the one at which the storage facilities are not overloaded. The importance of composition was mentioned in the comments on effect of type of collection. The more readily decomposable (biologically unstable, putrescible) the waste, the more frequently it should be collected. With the exception of fly production, putrescibility overrides all the other factors in the determination of frequency.

Frequency of collection probably has an effect on rate of waste generation, even though theoretically it would seem that a "steady-state" rate of discard should be reached that would be dependent upon income, standard of living, and other factors. One would expect an initial increase in amount of wastes discarded by the householders to be the immediate aftermath of the initiation of an increase in frequency, due to the discard of wastes accumulated during the time of the less frequent collection. However, after the backlog is eliminated, the rate of generation should drop to that which prevailed prior to the increase in frequency. This expectation was not met in an early study in California[5] in which it was found that with once-a-week collection, the weekly output per service was 14.5 kg (0.68 kg per capita-day), whereas with twice-a-week collection, the weekly output per service was 21.8 kg (1 kg per capita-day). Although the survey took place in the early 1950s, the relationship probably continues to hold true.

Certainly, the cost of collection service rises in proportion to the increase in frequency of collection, if for no other reason than that the distance and man-hours per household is a multiple of the number of times the refuse is collected per household. The California study indicated a 1.55 increase in man-hours per metric ton with twice-a-week collection. Moreover, the magnifying effect on amount of generation apparently exerted by an increase in frequency would result in more wastes to dispose, and, hence, an increase in costs. However, the increase in cost would be cancelled if the waste contained a recoverable and marketable resource.

The final factor to be considered, namely, effect on fly generation, perhaps surpasses the preceding ones in importance because of its bearing on public health. The relation between fly generation and frequency of collection rests upon the interruption of the fly life cycle accomplished by the imposition of an appropriate frequency. On the average, the fly completes its life cycle (egg to adult) in about 5 days. (Deviations from that time period may result from changes in climatic conditions.) Therefore, if the wastes are collected twice each week, the fly cycle is interrupted and fly generation is accordingly reduced. Another mitigating effect of an increased frequency is the reduction of the opportunities for flies to oviposit. The fewer the eggs, the less the resulting fly population.

3. Point of Collection

"Point of collection" refers to the position of the storage receptacle at the time of collection. The various possible positions are divided into two broad groups, namely curb and "backyard". Curbside collection may be regarded as being synonymous with "alley" collection since both technically take place entirely off the householder's property and the containers are handy to the collection vehicle. "Backyard" applies to any location on the householder's premises and need not necessarily refer to the rear half of his property.

The advantages and disadvantages of curbside collection are readily apparent. Advantages are mainly economic in nature in that the collection time per metric ton (man-min/Mg) is reduced. Referring again to the California study, the pickup time with 100% curb collection was found to average 100 man-min/Mg, whereas with 100% "backyard" collection, it averaged 165 man-min/Mg. The increase is not surprising in that on the average, about 30% of a collector's time is spent in walking on private property in backyard collection.

The disadvantages of curb collection are mainly in terms of adverse environmental impact. (By the same token, the disadvantages of curb collection redound to the advantage of "backyard" collection.) With curb collection, the refuse containers and their unsightly and malodorous contents are highly evident to passersby on the day of collection. Moreover, they are readily accessible to pets and to children. Of course, these adverse aspects can be ameliorated somewhat by the institution of a rigidly followed collection schedule, and of a regulation to the effect that emptied containers be removed from the curb within a reasonable period of time.

4. Pickup Density

Pickup density is a term given the numer of services per kilometer. Intuitively, one would surmise that the man-min/Mg of refuse would become fewer as the pickup density increased up to a certain point, because less time would be spent in traveling from one service to the next. In the California study,[5] the breakpoint in terms of pickup time occurred at a density of 15 services per kilometer, in that the man-min/Mg remained practically unchanged over the range of 15 to 107 service per kilometer. The increase in man-minutes became especially sharp as the density dropped below 15 services per kilometer. Thus, at six services per kilometer, the pickup time was twice that at 24 services per kilometer. Consequently, as the pickup density drops below ten or so per kilometer, the feasibility of instituting the rural container system becomes more apparent.[6]

5. Programming

Programming is a complex operation even for a medium-sized community. It involves planning routes and usage of manpower and equipment. The two disciplines, systems analysis and operations research, are of special utility in programming. Operations research is a process of evaluating various ways of using men and machines to find the most efficient arrangement. This aspect of collection was the center of a number of studies in the early 1970s.[7-10] The requirements for sound programming are a collection of accurate data, a firm comprehension of local conditions, and a thorough appreciation of the human element, as for example, of the work force. Among the fixed factors to consider in programming, i.e., factors not subject to change, are type of refuse, the method of disposal, the extent of resource recovery in progress or planned, the physical layout, and the climatic conditions, especially the meteorologic conditions.

6. Equipment

Generally, environmental and public health demand that the collection vehicle be covered and water tight. For the most efficient utilization, the capacity of the vehicle should be the maximum allowed by street width and traffic conditions. Type of waste dictates type of vehicle. For refuse in general, the conventional packer truck is appropriate. The truck can be modified slightly by adding on a rack for holding bundles of newspapers and a small bin to hold aluminum containers. Such modifications interfere only slightly with the operation of the vehicle. Open bed trucks may be used to haul demolition debris, tree trimmings, and other aesthetically innocuous materials.

Within the past decade, modifications of truck design have been tried that were aimed at minimizing the dependence upon manpower. As such the modifications took the form of automating the picking up and dumping of containers. Thus, one design involves positioning a conveyor belt and scoop such that the refuse container is picked up, emptied, and returned to the curb, all without human intervention. Another approach was to equip the truck with a hydraulic boom mechanism that can grasp a container placed on a curb, raise it, and dump the container's contents into the body of the truck.[11] Attempts at further automating collection through changes in conventional packer truck design have met with mixed success. A grave problem is that posed by parked automobiles. All of the adaptations presuppose an unimpeded access to the refuse containers. Moreover, they require the use of containers the capacity of which is greater than the (30 gal) of the conventional container ("garbage can").

IV. TRANSPORT (HAUL)

A. Introduction

A prime factor in the economics of collecting wastes is the cost of moving them from the point of generation to that of disposal or processing. Factors constitutive of this prime factor are equipment, fuel consumption, and manpower costs. The magnitude of the individual costs ultimately is one of mileage, or perhaps of time. The consequence of the distance factor is that a definite distance exists beyond which the transport of refuse in the collection vehicle becomes economically unfeasible. The time element enters in because the crew, excepting the driver, remains idle during the time of transport. Moreover, wear and tear on the collection vehicle is more a function of time while in motion rather than of distance covered. An exception, of course, is tire wear, which is solely a function of distance traveled. Time certainly is more inclusive than distance alone, because of the fact that the vehicle is not necessarily continuously in motion nor is the movement at a constant velocity. Consequently, the combined effects of time and distance should be taken into consideration in the determination of an economically practical length of haul with the collection vehicle.

What is to be done when the distance between the last service stop and the disposal site exceeds that economically practical? One approach, and it is the one conventionally followed at present, is to transfer the refuse from the collection truck to a vehicle of a larger capacity ("transfer truck") manned by a single individual, the driver. Where water transport is possible, a barge can take the place of a transfer truck. Rail transport would be an excellent substitute for truck transport were it not for the seeming impossibility for the railroad operators and the agencies responsible for moving municipal solid waste to reach mutually agreeable terms. Because of this difficulty, rail haul of municipal wastes in the U.S. is all but nonexistent.

The transfer of wastes from collection truck to transfer truck may be direct and involve only the use of a ramp; or it may be less direct in that the wastes are dumped into a receiving hopper or storage pit from which it is loaded onto the transfer truck. Not surprisingly, the facility at which the transfer is made is known as a "transfer station". Probably the best design for a transfer station is one which calls for a three-level installation. In the design, bays for unloading the collection trucks are installed at the top level. The middle level serves as the receiving pit, and has discharge hoppers installed at one end. The hoppers discharge into transfer trucks positioned at the lowermost level. The receiving pit is designed such that a bulldozer can push wastes directly into the hoppers which feed the transfer trucks.

For resource recovery, the transfer station should be the site of initial processing, usually in the form of separating potentially useful components from the nonuseful. Examples are recovery of paper fibers and of metals (containers). When a market for the products exists, such a removal becomes a fairly common practice.

The interrelationship between haul distance and time and costs of haul by collection vehicle and by transfer truck is clearly illustrated by the three curves in Figure 1. The curve for transfer station plus long haul does not begin at zero because regardless of distance of haul, the cost of the transfer operation remains constant, as is indicated in the figure. As the figure shows, a "crossover" point exists at which direct haul costs begin to exceed those of the combination of transfer station and long haul. The point at which the two curves cross is a function of several factors, the more important of which are labor costs (crew size and hourly wage), fuel costs, and traffic flow. Obviously, as values for any one or of all three go up, the shorter becomes the economically permissible distance and time.

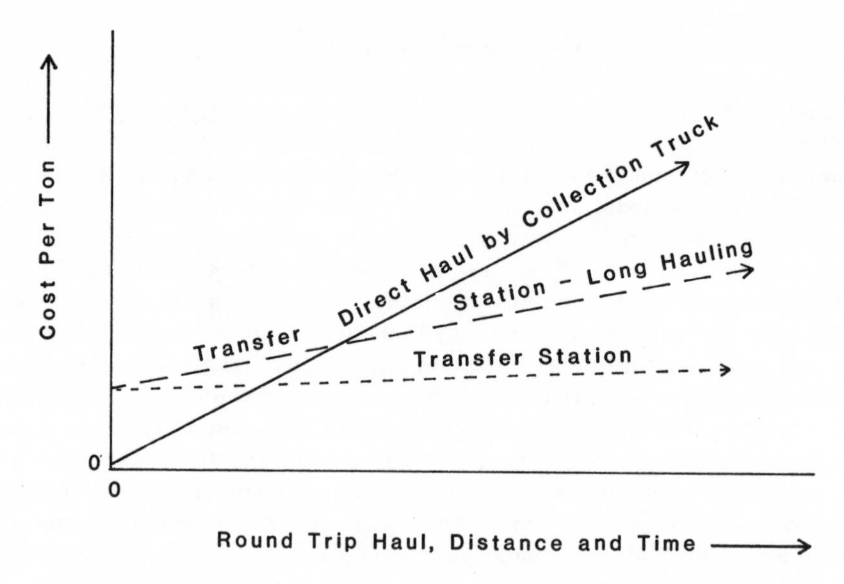

FIGURE 1. Relation of direct haul costs to those of a transfer station-long haul combination.

In the to-be-hoped-for event that railroad operators and the managers of solid wastes come to an agreement, certain requirements must be met to ensure the continued success of any arrangement established by them. The first requirement, of course, is that the rail transport be less expensive than truck transport. According to some knowledgeable individuals, rail transport would be cheaper at distances greater than 80 km. A second requirement is that an alternative arrangement be available in case the railroad employees go "on strike". In the absence of such an arrangement, the resulting accumulation of wastes could be disastrous in its impact. A third requirement pertains to the railroad cars in which the wastes are transported. The cars should be in good condition and designed to isolate the wastes from the external environment. Although not critical, a fourth one could be that the solid waste exporter own the rolling stock used to transport the wastes. A fifth requirement is the existence of a long-term and guaranteed performance contract that preferably is bonded. A sixth requirement is a quite obvious one, namely that the authorities and citizenry at the discharge end of the line agree to allow the wastes to be discharged in their sectors.

B. Pipe Transport

A solution to the many collection and transport problems would be had if an economically feasible means were found for transporting the municipal wastes through a pipeline from residences and other sites of waste generation to the disposal or resource recovery site. Traffic would no longer be an obstacle, collection would cease to be labor intensive, and the specter of a fuel shortage would disappear. Unfortunately, before this rather euphoric state of affairs can come to pass, a major obstacle must be hurdled, namely the high costs involved.

Although the technology for the pipe transport of solids is fairly well developed, it is not necessarily directly transferable to that of municipal refuse. Two types of pipe transport are available, namely hydraulic and pneumatic. As the term implies, in hydraulic transport the solids are moved while in suspension, usually as an aqueous suspension. The solids remain dry and are suspended in an air stream in pneumatic transport. In current practice, solids to be transported hydraulically (e.g., pulverized coal), characteristically have a uniform particle size and are granular in nature. In pneumatic transport the maximum allowable dimension of the particles is that of the inside diameter of the pipe, and maximum permissible density of the solids to be moved is a function of the velocity of the air stream.

One of the better analyses on hydraulic transport of wastes was made by Albrecht and Oberacker.[12] With respect to the economics involved, they estimated that the disposal costs (1975 dollars) with a wet transport system would range from $114 to $142/Mg, exclusive of the costs of the grinder system, which would add another $35 to $38. On the other hand, with conventional transport, the total cost for disposal at the time was only $39/Mg. They further state that before sewer transport is adopted on an appreciable scale, three aspects should receive further investigation, namely: (1) degree of sewer maintenance required; (2) compatibility of treatment systems with the heavier loadings caused by refuse; and (3) the extent to which modifications would have to be made to household plumbing and existing sewer lines.

Pneumatic systems are best used in situations characterized by highly dense populations such as are encountered in high-rise apartment complexes and in large institutions. An outstanding example of a pneumatic system application is one installed at Disney World in Florida.

In a typical pneumatic installation, vertical pipes connected to a horizontal transport pipe serve as feed inlets. In a high-rise apartment building, for instance, the vertical pipe takes the place of the conventional "garbage chute". At the junction between a vertical and the horizontal pipe is interposed a gate valve. The gate valves are programmed to open in series to release wastes from the vertical pipes into the horizontal pipe. After introduction into the horizontal pipe, the waste particles are moved pneumatically to a central collection chamber.

While the technology of pneumatic systems is fairly well developed, the costs are impressive. Capital costs are within a range of $800 to $1400 per apartment.

V. SOURCE SEPARATION

A. Introduction

"Source separation" in the common understanding of the term applies to the separation of solid waste at the point of generation into recyclable and nonrecyclable components. Thus, residential wastes would be sorted by the householder. The term often is extended to include the separation done at the small recycling centers or stations. The recyclables are collected separately with the intent of selling them for reuse. At present, the main components thus separated are paper, containers (cans), and glass (usually bottles).

Intuitively one would judge that it would be easier to collect and process a given discarded resource that had been kept separate from other wastes from the point of discard to the point of processing, than one that had been dumped directly into the general waste stream. While the basis for the intuitive conclusion is seemingly obvious, its soundness in a "real-life" situation is the subject of much debate between those in favor of source separation and those against it.

The procedure in source separation is for the generator to segregate his waste output into two or more fractions, the number and identity of which are based on the categories to be recovered. Logically, the separation is maintained throughout the subsequent steps in solid waste management. On the other hand, in centralized separation, the generator dumps all wastes into a single collection unit. His wastes are subsequently mixed with wastes gathered at other generation points, and the entire collected mass is brought to a central point for processing. It is only at the point of processing that the desired categories are separated one from the other and from the remaining collection of wastes.

B. Arguments For and Against Source Separation

The proponents of centralized separation hold that with the automation technology already available, a satisfactory degree of separation can be attained. It is their contention that the main advantage accruing from centralized separation as compared to source separation is the potential of procuring a reliable separation at a lower expenditure of energy and cash. They back their position by pointing to the alleged technological and human obstacles that supposedly beset source separation. As an example of a technological obstacle, they cite an apparent (i.e., to them) lack of equipment specifically designed to maintain the integrity of the separated categories during transport from the waste generator to the disposal or processing site. According to them, the human problems originate partly in the absence of a strong motivation and partly in a confusion as to classification identities.

The proponents of source separation counter with the argument that the energy expended in source separation is mainly that spent by the generator in dumping a few selected items in separate containers. The costs involved are accordingly low. Obstacles in the form of a supposedly reluctant cooperation and confusion on the part of the citizenry can be eliminated by a well thought-out and energetically conducted program of public education. Case histories of successful undertakings are cited to prove their point.[13]

One of the problems in source separation is the limitation placed by the technological nature of source separation upon the number and identity of the categories of waste resources that can be separated. The technological limitation comes from the need to collect and transport each category of items to its eventual destination, and yet maintain the separation throughout. This need places an upper limit on the number of separations that are technologically and economically feasible. Of course, in a real situation, the actual number of categories is decided by the number of those items for which there is a market. However, even if a market were available for each and every category of recoverable resource, the number of categories that could be feasibly accommodated would continue to be limited.

In terms of storage, the need to provide a container for each category of material is an expense in itself in that the number of containers is increased. Certainly, the cost of a single container large enough to hold the entire daily or weekly waste output is less than that of two or more containers to accommodate the output when split into two or more fractions. On the other hand, the additional cost would be minimized through the use of a recyclable containment material (e.g., burlap bags) for the separated recyclables.

Problems already afflicting collection systems could be multiplied by the need to maintain the separation of the sorted items during transport and discharge. However, compartmentalized vehicles designed to maintain the separation are available and are being used successfully. Of course, another approach is to collect the segregated recyclables in separate collection vehicles designed specifically for the purpose. Because of the nature of the recyclables and differences in rates of generation, the frequency of collection of the recyclables could be much less than that of unsegregated refuse which is contaminated with garbage.

An example of a compartmentalized system is the Mehrkammer-Mullsystem® manufactured by the Dornier System GmbH of Germany. The system has two complementary components, namely, a two-compartment container for use by the householder and a two-compartment packer truck. The container and truck are designed such that the contents of the container are automatically discharged into the

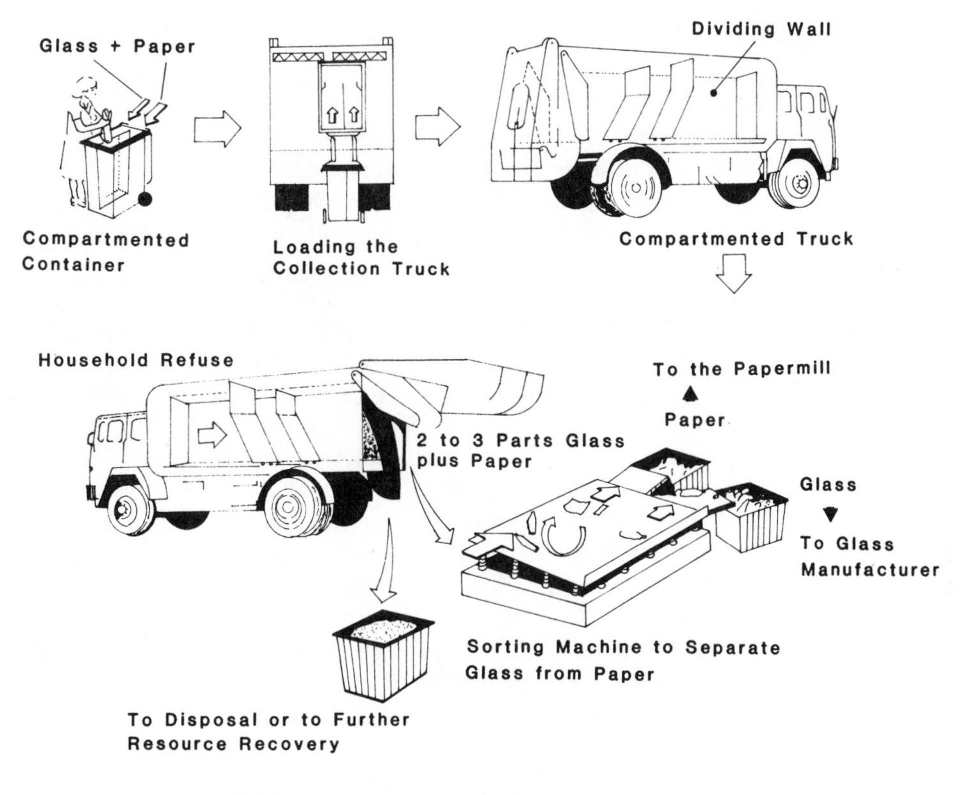

FIGURE 2. Truck designed for separate collection.

truck. At the discharge point, the compartments are unloaded simultaneously. The system and its operation are diagrammed in Figure 2.

The fact that there is a basis for the conviction had by source separation proponents to the effect that the motivation problem can be remedied by intensive public education, is attested by the successful wastepaper and beverage container drives that have been carried on in this country. However, the existence of an intractable minority of the populace resistant to cooperation of any sort can jeopardize an otherwise successful program. Despite the small number of individuals involved, the amount of waste discharged by this minority without sorting may be enough to seriously contaminate separated components. For example, flattened cans or garbage-tainted paper concealed in a bale of newsprint could be enough to render the bale unacceptable to the buyer. A possible solution for the problem would be to exempt the recalcitrants from the program and charge them for the privilege.

The confusion angle stems from the fact that many items in the waste stream do not fit neatly into a single category, and indeed, often may fit in two or more categories. Regardless of his or her confusion, the householder will place every item in some container. The problem arising from that fact is that in case of doubt, the choice will not be uniform among a given body of householders.

The economics of source separation are determined partially by factors common to any resource recovery activity, and partly peculiar to source separation. The important factor common to all resource recovery is in the form of the revenues and disposal cost savings associated with materials recovery and the amount of materials thus collected. Factors peculiar to source separation are the costs of collecting, publicizing, and administering the source separation program. The volume of material depends upon the extent of resident participation. The overall factor is the market price of the recovered materials.

In conclusion, centralized separation and source separation are not necessarily incompatible. Under certain conditions, a resource recovery program could be designed in which one would be complemented by the other.

Although they have nothing to offer in terms of technological information, two publications available at the time of this writing are outstanding in their informational content on methods and procedures for developing and implementing a program of source separation. They are "Source Separation and Recycling Program Implementation Manual" by John Madole of the Minnesota Pollution Control Agency,[14] and "Source Separation: The Community Awareness Program in Somerville and Marblehead, Massachusetts.[15] Of the two, the manual is the more complete and detailed in that it presents step-by-step directions. Its special utility is in its guidelines for making an economic analysis of source separation. The analysis is presented complete with work sheets and appropriate formulas. Among the recovery options described in the "Manual," one deals with separation programs for which support is mandated by ordinances. In this section, the residents are directed to segregate at the curbside recyclable materials and common refuse. Voluntary source separation programs that follow the curbside pickup procedure constitute the second option. The third option is one in which citizens drop their recyclables off at a location where the recyclables are stored and processed for market. Applications of the first option are rare, while those of the second are mostly limited to newspaper salvage. Of the three options, the third is the one most commonly followed.

Although restricted in scope to the means of motivating and educating the public, the second publication ("Community Awareness Program") has its own utility in that it instructs the reader by means of describing two case histories. An especially valuable feature of the publication is its section on guidelines. The topics covered in the guidelines are the general steps to be followed in establishing community awareness. They are scheduling, securing local participation, clarification of roles, planning the program, and obtaining commercial support. Specific program components relative to the media are logotype and title, newspapers, the community letter, radio and cable television, commercial television, posters, graphic displays, stickers, calendars, and the school.

REFERENCES

1. **Tchobanoglous, G., Thiesen, H., and Eliassen, R.,** *Solid Wastes Engineering Principles and Management Issues*, McGraw-Hill, New York, 1977.
2. Municipal Refuse Disposal, Institute for Solid Wastes, American Public Works Association, Public Administration Service, Chicago, Ill., 1970.
3. **Wilson, D. G.,** Ed., *Handbook of Solid Waste Management*, D. Van Nostrand, New York, 1977, 752.
4. **Cimino, J. A.,** Health and safety in the solid waste industry, *Am. J. Publ. Heath*, 65(1), 38, 1975.
5. An Analysis of Refuse Collection and Sanitary Landfill Disposal, Tech. Bull. No. 8, Ser. 37, Sanitary Engineering Research Laboratory, University of California, Berkeley, December 1952.
6. **Reindl, J.,** Rural Solid Waste Collection Systems, G2853, Cooperative Extension Programs, University of Wisconsin Extension, Madison, Wis., March 1977.
7. **Clark, R. M. and Helma, B. P.,** Fleet selection for solid waste collection systems, *J. Sanit. Eng. Div. ASCE*, 97(SA-1), 71, 1972.
8. **Bodner, R. M., Cassell, E. A., and Andros, J. P.,** Optimal routing of refuse collection vehicles, *J. Sanit. Eng. Div. ASCE*, 96(SA4), 893, 1970.
9. **Liebman, J. C.,** Systems Approaches to Solid Waste Collection Problems, paper presented at the Meet. A.A.A.S., Chicago, Ill., December 1970.
10. **Truitt, M. M., Liebman, J. C., and Kruse, C. W.,** Mathematical Modeling of Solid Waste Collection Policies, report by John Hopkins University for Bureau of Solid Waste Management, U.S. Department of Health, Education, and Welfare, PHS, 1970.
11. **Stragier, M. G.,** Barrel-snatcher eliminates refuse collection employment problems, *Publ. Works*, 102(1), 64, 1971.
12. **Albrecht, O. W. and Oberacker, D. A.,** Sewer transport of household refuse—a replacement for the refuse truck, in *News Environ. Res. Cincinnati*, Solid and Hazardous Waste Research Laboratory, U.S. EPA, May 9, 1975, 4.
13. **Hill, S. A.,** Appropriate technology for resource recovery, *N.C.R.R. Bull.*, 11(1), 3, 1981.
14. **Madole, J.,** Source Separation and Recycling Program Implementation Manual (draft copy), Minnesota Pollution Control Agency, St. Paul, 1979.
15. Source Separation: The Community Awareness Program in Somerville and Marblehead, Massachusetts, EPA/530/SW-551, U.S. EPA, November 1976.

Chapter 3

PLANNING, DESIGNING, AND MODELING THE RESOURCE RECOVERY FACILITY

I. INTRODUCTION

A. Factors That Affect Quality of Design

Key requisites for the design of a complete resource recovery plant are decisions as to which materials and in which forms energy are to be recovered, and whether the two can be recovered in economically feasible amounts. The decision regarding materials can be made upon the basis of results of waste characterization and marketing studies. Through a characterization study, it is possible to ascertain amounts and types of potentially recoverable materials. Results of a market study establish the types and quantities of the materials that can be used or sold. In effect, the results of the market study can constitute a determinant in the economic feasibility of the recovery process.

B. Integration of Unit Processes

Generally, every resource recovery plant consists of a "front-end" and a "back-end" system. The two terms are fairly self-explanatory. The front-end system, usually composed of one or more subsystems, provides the processing through which a feedstock is prepared for the back-end system or systems. At the time of this writing no great diversity existed with respect to front-end configuration, in that almost all front-end systems relied upon shredding as the basic processing element. However, a front-end system need not be confined basically to shredding. It may include other unit operations such as screening, air classification, and magnetic separation. Back-end systems are loosely defined in the resource recovery industry as systems or subsystems designed to treat certain concentrated components of municipal solid waste (MSW). Therefore, the functioning of back-end systems depends upon some form of prior processing, in the course of which certain materials in the feedstock are separated from the waste stream and concentrated (i.e., through front-end processing) before being introduced into the back-end system. The air classified light and heavy fractions of MSW are two examples of feedstocks that have been separated and concentrated for back-end systems. The light fraction is a feedstock suitable for an energy recovery back-end system. The heavy fraction can serve as a feedstock in magnetic and glass recovery back-end systems.

Because front-end systems typically are similar in general lay-out, the development of a plant design usually begins with the establishment of a front-end system capable of accommodating the type and quantity of material to be processed. At the same time, however, the requirements of the back-end systems must be kept in mind, because ultimately the front-end and back-end systems must be integrated.

C. Mass Balances

The integration of front-end and back-end systems—and for that matter, of all the unit processes—should be preceded by the development of a detailed mass balance. Because the need for a mass balance seems self-evident, one would expect that the usual practice would be to make one as a routine step in plant design. The fact is, however, that more than a few resource recovery plants in operation today are handicapped by design deficiencies traceable directly to a reliance upon mass balances that are poorly defined with respect to the processing lines. The unfortunate result is that the processing equipment is undersized in some plants and oversized in others. In other cases, because of a faulty mass balance, predicted separation or splits of material at or by a given piece of equipment are not at the predicted levels. The consequence is the "starving" (underburdening) of one downstream process and the overburdening of another.

D. Knowledge of Feedstock, Product, and Equipment

In the main, faulty mass balances can be traced to two causes: (1) an incomplete knowledge of the components and properties of MSW; and (2) an inadequate conceptual grasp of the operation and performance of the unit processes involved in the functioning of the recovery facility. A classic manifestation of the two inadequacies is the stipulation that the material passed through an air classifier have a given percentage (e.g., 85%) of material reporting in the light fraction, without simultaneously specifying the composition and moisture content of the material being processed. The failure to so specify results in the ignoring of two factors, namely, composition and moisture content, that directly influence the mass split in an air classifier under a given set of operating conditions.

Another cause of design flaws is the general unavailability of performance data for various pieces of equipment. For example, data on the relationship between product particle size and throughout capacity of size reduction units are virtually nonexistent. Because of the lack of such data, there is the chance that a plant manager may find that when he changes a shredder's grates to produce a smaller particle size, the throughput capacity of his shredder is reduced such that the designed throughput rate cannot be met. The consequence, as far as the overall operation is concerned, is the operation of downstream equipment at less than designed capacity.

The two causes mentioned in the opening paragraph, namely, incomplete knowledge and inadequate conceptual grasp, can lead to the aggravation of other difficulties. In the past, incomplete knowledge, or to express it bluntly, ignorance of the properties of MSW, has been the more responsible of the two for magnification of the many problems inherent in the operation of a resource recovery plant. This is especially true with respect to the storage of shredded refuse.

After a prolonged history of the many problems encountered in the storage of shredded refuse, individuals responsible for the design of plants, equipment, and operation in the solid waste industry finally are beginning to recognize the fact that those problems arose ultimately from the failure to realize that shredded solid waste acts like a compressible fluid. For example, if shredded refuse is piled to a height (or depth) of 6 to 9 m and is allowed to remain undisturbed, material at the bottom of the pile eventually reaches densities ranging from 320 to as much as 640kg/m^3 (20 to 40 lb/ft^3). The densification is due both to the onset of biological decomposition processes and to the weight of the overlying mass of material. The compression can be so great that drastic measures—even the use of dynamite—may be required to loosen ad move shredded refuse that has been piled high and allowed to remain undisturbed for a couple of weeks inside a container. Needless to say, in the absence of prior experience with the problem, basic experimentation to determine the pertinent properties of shredded waste would have alerted designers to this storage problem.

E. Manner of Recompense

Factors that lead to poor design directly or indirectly are not all equipment, feedstock, or product oriented. Thus, economics may constitute a factor by way of the manner in which the operator of a recovery plant is recompensed for material produced by his plant. He may be paid solely on the basis of quantity of production or his pay may depend upon quantity plus quality. Although of the two one would feel intuitively that the second basis is to be preferred, the fact remains that the preferred choice is not always the one that is made. This especially holds true in RDF production, in which compensation often is on the basis of quantity (in tonnage, or in heat units) produced rather than on quantity *together with* quality. Using quantity as a sole basis results in an overtaxing of downstream equipment so as to maximize production, often at the expense of quality.

F. Equipment Selection

Equipment must be matched to the properties of the materials to be processed, to the throughputs required, and to the desired specifications of the product. Secondary considerations generally involve space limitations, which may dictate the adoption of one equipment configuration rather than another. Because of the many problems rising from materials handling, spillage, dust generation, undersized equipment, and pronounced wear-and-tear in resource recovery plants, the two considerations, repair and housekeeping, should be near the top of the list of priorities in the selection of equipment. Almost any manager of a resource recovery plant would agree that if the preceding advice had been followed in the equipping of his plant, his plant now would have lower maintenance costs and a substantially greater production. Consequently, the following measures take on a significant meaning after the plant has been built and is in operation: the ensuring of a ready access to mechanical and electrical equipment; the provision of adequate walkways to, on, and around equipment; a limitation on the number of long stairways and ladders; the making of provisions for maintenance and repair of equipment; and the having on hand a properly equipped machine shop, and welding and maintenance facilities. If these items are satisfactorily addressed in the design stage, time and money will be saved because of less downtime for repair and maintenance after the plant is placed in operation.

With respect to the overall design of a resource recovery facility, a horizontal layout of equipment is preferable to a vertical layout. With a horizontal layout, maintenance problems associated with large pieces of equipment installed high above the main floor level are avoided. Furthermore, if it should become necessary to remove a large piece of equipment, the removal can be accomplished with greater ease.

II. STORAGE AT THE PROCESSING SITE

A. The Case Against Storage

The troubled experience had to date with the storage of shredded refuse and RDF at the processing site is a convincing argument for making every effort to bypass the step. The major reason for the trouble is in the design of existing storage hoppers and bins. They are not designed to permit a satisfactory and reliable removal of shredded refuse that has a relatively large paper content (i.e., 30 to 60% on an as-processed basis), has a moisture content of 30 to 45% (oven-dry basis), and has not been disturbed over a period of 10 hr or longer.

As a rule of thumb, the tendency of shredded refuse to compact and remain "set" becomes more pronounced as the concentration of cellulose is increased until it reaches 60%. At higher concentrations the effect remains the same as at 60%. The tendency is comparable with respect to moisture content, excepting that it is at its maximum when the moisture content is 40% and higher. These latter two concentrations are the ones generally encountered in practice. The noncellulosic fraction of shredded MSW is more amenable than the cellulosic fraction to storage and removal. Grit, glass, and metallic particles in the noncellulosic fraction apparently impart a degree of looseness to the material, thereby lessening its tendency to compact and remain "set".

Other problems associated with storage systems take the form of a more or less severe abrasion of floors and of removal hardware installed within the hoppers or bins, and of the difficulties encountered in materials handling. The difficulties alluded to here are those that come from the tendency of the stored material "to bridge" and from the jamming of the removal equipment by the material being moved. The latter leads to an overtaxing of the drive system that powers the removal equipment.

B. Solutions for Storage Problems

Short of keeping the pile at a shallow depth, the most trouble-free storage system under present circumstances is one that involves the use of a large bin and the removal of the stored material by means of a bulldozer or front-end loader. The shallow pile approach is impractical because of the inordinately large floor space required. A response to the second approach is the large door for "access" now being provided in a number of commercially available storage systems. In the absence of an access door, plant personnel usually will install one as soon as it becomes necessary to use a heavy-duty item (e.g., front-end loader) to remove material that has become compacted through storage. Another approach is to store MSW in its raw rather than in its shredded state. Raw MSW has been piled to depths of 6 to 8 m, and as such was stored for weeks, and yet could be readily removed and loaded by a front-end loader. However, it should be kept in mind that on-site storage of raw MSW is not appropriate for all applications, and in many locations is prohibited by rules and regulations. Recommended schemes for storage and removal of shredded cellulosic wastes are listed in the order of their suitability in Table 1.

The section on storage and removal would not be complete without mention of two design approaches regarding material storage and removal, namely, the so-called "first-in/first-out" and "first-in/last-out" approaches. In the first approach, material that has been stored the longest is the first to be removed, whereas in the second approach, freshly deposited material is the first to be removed. With the first-in/last-out approach, it is possible that some material would remain in the storage bin for a long time. However, it should be noted that if the storage period is significantly long, the first approach becomes in effect a first-in/last-out approach. This fact should be kept in mind if a storage period longer than 10 hr is anticipated.

III. BELT CONVEYORS

In a resource recovery facility, the transfer of material from one unit to another is accomplished most often by means of belt conveyors. Consequently, the provision of properly designed and specified conveyors is essential to the efficient functioning of the entire facility, especially since a number of conveyors usually is involved. Proper design and specifications include not only those which are standard for any reliable conveyor, but also certain others that are related to the nature of the various refuse fractions typically handled in a resource recovery facility. The first of the latter specifications refers to the maximum permissible angle of inclination of the conveyor. A list of inclinations recommended for various particle sizes and densities is given in Table 2. At inclinations steeper than those suggested in the table, the conveyor belt should be cleated since the material is likely to roll or slip back down the belt. It should be noted that the moisture content of the processed material has a significant influence on the tendency of the material to slip on a noncleated belt surface. The lower the moisture content, the greater is the tendency to slip. A second specification is that belted conveyors be of the troughed type, or be rubber-skirted along their entire length. The use of a troughed or skirted belt minimizes spillage over the edge of the belt. An

Table 1
RECOMMENDED SCHEMES FOR STORING AND REMOVING THE SHREDDED CELLULOSIC FRACTION OF MSW[a]

Do not store shredded cellulosic fractions of MSW
Store raw MSW instead of the cellulosic fraction
Store shredded cellulosic fractions of MSW such that the depth of the material is less than 2m
If depths are greater than 2 m, remove the material in less than 10 hr
If depths are greater than 2 m and storage times exceed 10 hr, use a bulldozer or front-end loader to remove the material from storage

Note: For the purposes of this table the term "shredded cellulosic fractions" of MSW includes shredded raw MSW with or without magnetic metals removed and air classified light fraction (RDF), and excludes screened air classified light fraction, i.e., trommel oversize material.

[a]Listed in order of suitability.

Table 2
RECOMMENDED MAXIMUM ANGLE OF INCLINATION FOR NONCLEATED FLAT AND TROUGH BELT CONVEYORS[a]

	Bulk density		Angle of inclination from the horizontal
	lb/ft^3	kg/m^3	
Raw MSW	10—14	160—224	25
Shredded MSW (25 to 5 cm nominal size)	9—13	144—208	25
Air classified lights (25 to 5 cm nominal size)	4—7	64—112	20
Screened air classified lights (trommel oversize, 15 to 5 cm nominal size)	2—4	32—64	15
Shredded ferrous scrap (10 to 2.5 cm nominal size)	50—65	80—104	20
Pretrommel oversize (25 to 5 cm nominal size)	4—6	64—96	20
Pretrommel undersize (12.5 to 10 cm nominal size)	15—25	240—400	22
RDF pellets (2.5 to 1.25 cm pellet diameter)	25—40	400—640	15

[a]Based on as yet unreported experience had by the authors.

additional specification applies to conveyors installed to move refuse fractions containing dirt, glass fines, or fine cellulosic fibers. Conveyors used for the latter purpose should be equipped with full-length dust covers designed to allow ready access to the shielded belt. The provision for ready access can be made through the use of either bolted or clamped covers. Moreover, all covered conveyors carrying dusty materials should be equipped with a dust collection system adequately sized to collect and remove the dust. Generally, a dust collection system consists of a blower, ducting, and either a cyclone or a baghouse, or both. Conveyors used for transporting shredded MSW should be provided with a water spray system if a fire hazard (i.e., from a shredder explosion) should exist.

A cleaning and collection apparatus should be installed at the head pulley to prevent spillage and carry-back of material. The apparatus may be one of any number of belt wiper arrangements. However, it should be pointed out that none of the presently available belt wiper devices are completely effective in the removal of material adhering to the belt surface. Nevertheless, the units do limit or may even entirely prevent the carry-back of the material. Prevention of the carry-back is important because the material is dislodged from the belt upon its return as it (the belt) passes over the return idlers. The dislodged material eventually finds its way into the conveyor superstructure or onto the floor. For ease of repair and maintenance, the direction of travel of the conveyor belts should be readily reversible, either mechanically or electrically.

IV. FLOW-RATE CONTROL EQUIPMENT

Inasmuch as at the time of this writing, loading by means of a front-end loader was the best available method of controlling the feed rate into the front-end systems, the provision of a control system for monitoring a uniform flow rate on a weight basis was a function of the expertise and effort of the operator of the front-end loader. A conscientious equipment operator generally can maintain the flow of refuse into a conveyor within a factor of 50% of the normal desired throughput. Thus, if the desired throughput were 30 Mg/hr, a well trained loader operator could keep the feed rate within a range of 15 to 45 Mg/hr, perhaps for as long as an hour at a time.

A mechanical means of controlling flow rate involves the imposition of "on" and "off" sequences. However, it is an unsatisfactory means because it may lead to an overheating of the electric motors. More significant in terms of uniformity of flow rate, it results in a poorly modulated throughput.

Because of its present inadequacies, the control of the flow rate of refuse into and through a processing system is perhaps the most fundamental area of refuse processing that requires an immediate and concerted research and developmental effort.

V. PNEUMATIC TRANSPORT

Because pneumatic transport phenomena have an extremely important position in resource recovery processing, a separate section is devoted solely to a discussion of them and of their influence. Pneumatic transport probably is second only to mechanical conveyance as the most widely used generic system in resource recovery processing.

The term "pneumatic transport system" is used in its generic sense in this book, and as such, refers to such processes as air transport, air fluidization, and cyclone separation of air and solids. These and other like processes are major features of the various recovery subsystems, two examples of which are air classification and the pneumatic transport of RDF. In this section, the discussion is restricted to the pneumatic conveyance of shredded MSW, especially of the air classified light fraction and of posttrommel oversize material (screened light fraction).

Because of the physical nature of shredded solid waste, the design of a pneumatic system to convey it becomes a difficult and perplexing task. For instance, MSW is inherently very abrasive because of the assortment of small glass, dirt, and other particles entrained in it. To complicate the problem even further, shredded MSW is very compressible. As such, in a transport line it tends to attach to and compact against transition points and at points of change in direction of flow, as for example, at elbows and "tees". These and other problems encountered in pneumatic transport are as old as the resource recovery industry itself. Two especially discouraging aspects of the situation are (1) the continuing prevalence of the types of equipment and processing systems demonstrated by past experience as being unsatisfactory; and (2) a pattern of using the same designs and techniques over and over again.

Particularly troublesome design problems are abrasion of fan blades and of the interior surface of material-handling blowers, and of abrasion and plugging of the transport line. The problems are due primarily to the abrasive nature of the fines entrained in shredded refuse transported by a high velocity air stream. Obviously, the use of wear-resistant duct work, piping, and blower internal parts is mandatory in the pneumatic transport of shredded refuse. Despite the seriousness of the problem, to date only one published study[1] had been addressed to the matter and to the utilization of wear-resistant material in the fashioning of equipment and parts for the transport of shredded refuse.

The accumulation of material within pneumatic pipelines and subsequent plugging has been a common experience at most refuse processing facilities. The causes of the problem have not been precisely identified, nor has the problem been adequately studied and reported. Suggested causes are abrupt changes in the direction of the flow of material, improper moisture content, presence of oversized or irregularly-shaped particles (e.g., wire, springs, long rags), an air velocity insufficient to fluidize the particles, and a wall temperature of tubes exposed to outdoor conditions that may be either too hot or too cold.

Thus far, published research detailing the interaction between a mixture of air and refuse particles with the walls of a pneumatic transport line is either extremely scarce or perhaps even nonexistent. Yet, basic research and publications thereof are needed to shed light on the mechanisms associated with wall wear, wall-particle cohesive forces, and duct blockage.

An important and practical lesson to be learned from the preceding discussion is that the blower should be positioned on the air-handling side of the cyclone collector or baghouse. Any additional pressure loss resulting from placing the blower on the downstream side of the cyclone or baghouse is more than offset by the reduction in wear problems accomplished by placing the blower upstream of the separation devices. Another, and possibly even more valuable lesson would be to the effect that until workable and economically feasible solutions are found to the many problems that beset pneumatic transport, a prudent plant designer would do well to rely upon mechanical conveyors (e.g., belt) wherever possible and to minimize dependence on pneumatic transport.

VI. MODELING OF SYSTEM MASS AND ENERGY BALANCES

The foundation of a rational design of a resource recovery plant is a detailed and comprehensive system mass balance broken down into each of the proposed plant's individual unit processes. Data from the resulting mass and energy balances coupled with a given set of material and energy specifications make it possible to properly size processing equipment (e.g., physical size, motor size, extent and magnitude of operating conditions), and subsequently size the overall plant (e.g., area of building floor, height of walls). Having specified the equipment and facility, one can calculate capital, operation and maintenance, and overall system costs. Finally, plant revenues can be calculated on the basis of the predicted flow rates and estimated unit selling prices of specified recovered materials (e.g., ferrous and aluminum scrap, RDF). In summary then, a mass balance is the basis upon which an orderly program can be developed to specify an entire resource recovery project—proceeding from a mass and energy balance on a single unit process, and continuing through the calculation of the operating costs and revenues for the entire resource recovery processing system.

A. Method of Establishing a Material and Energy Balance

The model described in the paragraphs which follow is based upon information obtained from a past study,[2] and incorporates basic elements of previously formulated system models.[3] The new model can be used in the simulation of mass and energy balances for a variety of front-end processing systems. The model differs conceptually from the previous versions and, in addition, involves a unique feature, namely, a recovery factor transfer function (RFTF) for each unit operation. With the use of the factor, a versatile model can be developed to describe the system mass balance. Thus, the RFTF is the key to the versatility of the model. At present, the factor can be adequately defined only upon the completion of a detailed analysis and interpretation of data collected from operating facilities. As time progresses, this complete dependence upon the empirical approach will be lessened, and eventually it will be possible to develop a recovery factor transfer function from basic principles. In some instances, recovery factors for certain processes can be calculated from data collected in past and in ongoing operations. Examples are data collected as a part of the St. Louis[4] and Ames[5] operations. Recovery factors for paper, plastic, ferrous metals, etc., that are based on field data collected during the testing of air classifiers in various parts of the U.S., are reported in two publications.[6,7]

B. The Recovery Factor Transform Function Matrix

Essentially, through use of a simple scalar matrix, the component mass inputs (e.g., Fe, Al, glass, paper, plastics) are transformed by way of the RFTF into the outputs for that operation. The intrinsic utility of this modeling approach is in the fact that the RFTF has a physical basis which can be established either through testing or from previously developed analytical expressions. The concept is illustrated by the following example in which is examined a binary separation device—in this case an air classifier. (A binary separation device, such as an air classifier, divides an input stream into two output streams.) If the input and output matrices to an air classifier are designated as U and X and Y, respectively, and are composed of n components of mass fractions u, x, and y, respectively, then the modeling situation is as shown in Figure 1. As mentioned previously, the RFTF (R_i in Figure 1) can be determined analytically. If adequate analytical expressions have not been developed, it can be determined through testing. The component recovery factors, obtained by way of any one of the aforementioned means, form the entries of RFTF matrix. It follows then, as shown in Figure 1, that the output streams and their mass fractions can be determined through the utilization of R and R′. Figure 1 also shows that by conservation of mass in a binary separator, $R + R' = 1$. Therefore, having established R, it is easy to calculate R′.

The recovery factor transfer function for each unit process can be expanded to include recovery factors for refuse components (on a dry weight basis) and component moisture content (free water). Consequently, the transfer functions account for solid and water separation, with the latter typically expressed as a moisture loss, (e.g., water vaporized during shredding). The reason for adding this degree of sophistication is the fact that ash content, and heating value, as well as a number of refuse component properties are usually reported on a dry weight basis. Consequently, to facilitate the computation of bulk stream properties from the summation of the properties of the components, it is convenient to deal with a recovery factor that operates on the dry weight of the solid components and their respective inherent water contents. The heating value of the light fraction is an example of a bulk stream property. The heating values of paper, plastic, etc. are examples of properties of the components.

A further sophistication of the modeling algorithms can be made through the application of fundamental governing equations. This approach generally requires a definition of the component size distributions (i.e., of Fe, Al, glass, paper, etc.) at each unit process, because just as it does in screening, particle size also typically governs the processes of separation in air classification. However, air classification differs somewhat in that it is governed by the size, geometry, and density of the material.

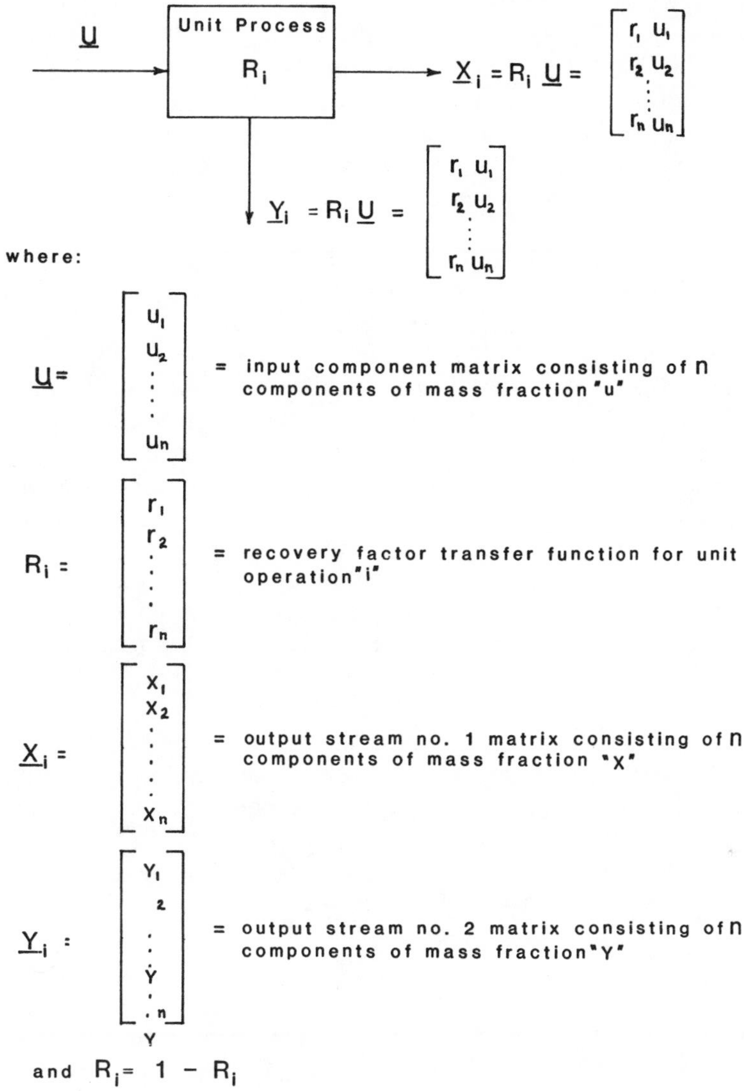

Unit Process R_i

$$\underline{U} \rightarrow R_i \rightarrow \underline{X}_i = R_i\,\underline{U} = \begin{bmatrix} r_1 u_1 \\ r_2 u_2 \\ \vdots \\ r_n u_n \end{bmatrix}$$

$$\underline{Y}_i = \bar{R}_i\,\underline{U} = \begin{bmatrix} \bar{r}_1 u_1 \\ \bar{r}_2 u_2 \\ \vdots \\ \bar{r}_n u_n \end{bmatrix}$$

where:

$$\underline{U} = \begin{bmatrix} u_1 \\ u_2 \\ \vdots \\ u_n \end{bmatrix}$$ = input component matrix consisting of n components of mass fraction "u"

$$R_i = \begin{bmatrix} r_1 \\ r_2 \\ \vdots \\ r_n \end{bmatrix}$$ = recovery factor transfer function for unit operation "i"

$$\underline{X}_i = \begin{bmatrix} x_1 \\ x_2 \\ \vdots \\ x_n \end{bmatrix}$$ = output stream no. 1 matrix consisting of n components of mass fraction "x"

$$\underline{Y}_i = \begin{bmatrix} y_1 \\ y_2 \\ \vdots \\ y_n \end{bmatrix}$$ = output stream no. 2 matrix consisting of n components of mass fraction "y"

and $\bar{R}_i = 1 - R_i$

FIGURE 1. Modeling algorithm for a mass balance of a unit process

The use of fundamental governing equations in the modeling of resource recovery processing is a significant departure from the current idea of system modeling had by industry. In the approach taken by industry, physical and mechanical properties of the components of refuse are not correlated with the effectiveness of the separation processes. The failure to make the correlation redounds to the detriment of any resulting modeling, because in practically every unit operation in a recovery facility, particle size and component densities are the primary material properties that govern the degree of separation of the various components.

However, the reasoning behind an analytical development of governing equations is given for academic purposes in the present discussion.

C. Rationale in the Development of the Model

Due to the nature of the separation devices typically used in resource recovery, the key to the development of a useful model for describing a particular system (e.g., RDF production) is a knowledge of the size distributions and other basic physical properties of the individual refuse components. Implicit in the preceding statement is an analytical description of raw and shredded MSW, since separation in most processing equipment is primarily a function of size and density of the throughput material. With the establishment of the size distributions of the components in the raw and in the shredded MSW, one can develop analytical expressions for describing the separation of components within the various separating systems. For example, in air classification, if the particle size distributions of the individual components of refuse after primary size reduction are given, it is possible through the use of a basic aerodynamic force balance to develop for each component a set of governing relations that describe the component mass fractions that will fly with the light fraction as a function of the average velocity of the air stream through the air classifier (see Volume I chapter titled "Air Classification"). The summation of the mass fractions of all the components flying with the light fraction provides the total mass fraction of the light fraction as a function of velocity. Furthermore, having established the mass balance on an analytical basis, bulk stream properties such as moisture content, ash content, and heating value can be calculated analytically through a summation of the products of the component mass fractions and their respective moisture contents, ash contents, and heating values.

Using air classification as an example, the preceding discussion and theory can be represented mathematically. The analytical expression for computing the mass of the fraction reporting to the light fraction is expressed in the form of:

$$m_{f_i} \mid_{LF} = a\, V^b\, G(x_i) \tag{1}$$

where $m_{f_i} \mid_{LF}$ is the mass fraction of component "i" in the light fraction; a and b are constants; V is the average air classifier column velocity; and $G(x_i)$is a mathematical function describing the input size distribution of component "i", e.g., Rosin-Rammler equation.

The mass of the fraction reporting to the heavy fraction is computed from the calculated value of the light fraction and by invoking the law of conservation of mass as follows:

$$m_{f_i} \mid_{HF} = 1 - m_{f_i} \mid_{LF} \tag{2}$$

where $m_{f_i} \mid_{HF}$ = mass fraction of component "i" in the heavy fraction. To account for an additive intrinsic property of the light fraction (P_{LF}), the following expression (i.e., algebraic summation) can be used:

$$P_{LF} = \sum_{i=1}^{n} m_{f_i} \mid_{LF} P_i \tag{3}$$

where P_{LF} is the property of the light fraction (e.g., moisture content, heating value, etc.); $m_{f_i} \mid_{LF}$ is the mass fraction of component "i" in the light fraction; p_i is the property of component "i"; and n is the number of refuse components.

The component recovery factors used in the transfer function for the unit process (e.g., air classification) can also be calculated analytically because the component recovery factor (RF_i) for a binary separation device can be represented simply by

$$RF_i = (m_{f_i} \mid_{LF})\,(m_{f_i} \mid_{HF} + m_{f_i} \mid_{LF}) \tag{4}$$

where the values of

$$m_{f_i} \mid_{LF} \text{ and } m_{f_i} \mid_{HF}$$

are given by analytical expressions such as Equations 1 and 2, respectively.

A mass balance can be similarly established for other separation devices, be they binary, tertiary, or other. Regardless of whether the values of the component recovery factors are determined analytically or empirically, once they are established for each unit process, the essential building blocks of the system model can be assembled. The models of each unit process is separate from the models of the other unit processes in a system, because the recovery factors are based upon size distributions of the refuse components, and where possible, upon other component physical properties as well. In other words, each model is a separate entity and complete unto itself. Consequently, in the system model, the unit operations can be arranged to evaluate different processing trains. The practicality and utility of modeling based on RFTFs is further enhanced if recovery factors are also used to provide the basis for specifying process performance. An example would be the specifying of the paper recovery in the light fraction as a percentage of the input paper.

Modeling techniques thus far reported in the literature are all dependent upon the interrelationships of unit operations, that is, upon their relative locations with respect to one another. The dependence is mostly a function of the neglect to take into consideration the size distributions of the components of the refuse in the input streams. Because of the dependence, the sequence of unit operations cannot be interchanged in earlier models without having to develop a completely new model. This need to develop multiple models of processing systems is eliminated through the use of recovery factors that account for component size distribution. Moreover, through such an approach, the utility of the resulting model is maximized.

In Figure 2 a block diagram in which the RFTF approach is followed and which represents a hypothetical RDF recovery system in which pretrommeling, shredding, ferrous separation, and air classification are used to produce RDF. If it were based upon the use of RFTFs, block diagram of any hypothetical unit process would resemble the diagram shown in Figure 1.

The RFTF modeling approach is applicable to all RDF recovery systems, and perhaps most importantly, it can be used to evaluate means for optimizing existing recovery systems. Examples of the latter are (1) the institution of changes in equipment operating parameters that result in changes in the recovery factor transfer functions; and (2) the use of additional process equipment. Another advantage resulting from the versatility of the modeling technique is a capability for evaluating and optimizing various hypothetical processing trains for RDF quality and yield. Here the term "hypothetical" refers to system models that have no actual plant to mimic. For example, as of this writing there is no RDF facility operating in the U.S. that utilizes the processing sequence shown in Figure 2.

D. Setting Up a System Mass Balance

A sample calculation of the mass balance of an RDF recovery system that does not incorporate pretrommeling is given in Table 3. The mass balance of each unit process in the system was calculated on the basis of an assumed initial waste composition. (The system is diagrammed in Figure 3.) Recovery factor matrices for each unit process are shown in the table. In Table 4 are listed the ash content calculations for each stream in the process; in Table 5, the heating value calculations; and in Table 6, the moisture content calculations.

The magnitude of the calculations required for the relatively simple system described in Tables 3 to 6 demonstrates the need for computer solutions when a more complicated system is involved. The linear algebraic equations needed for the method described herein lend themselves to a straightforward computer encoding.

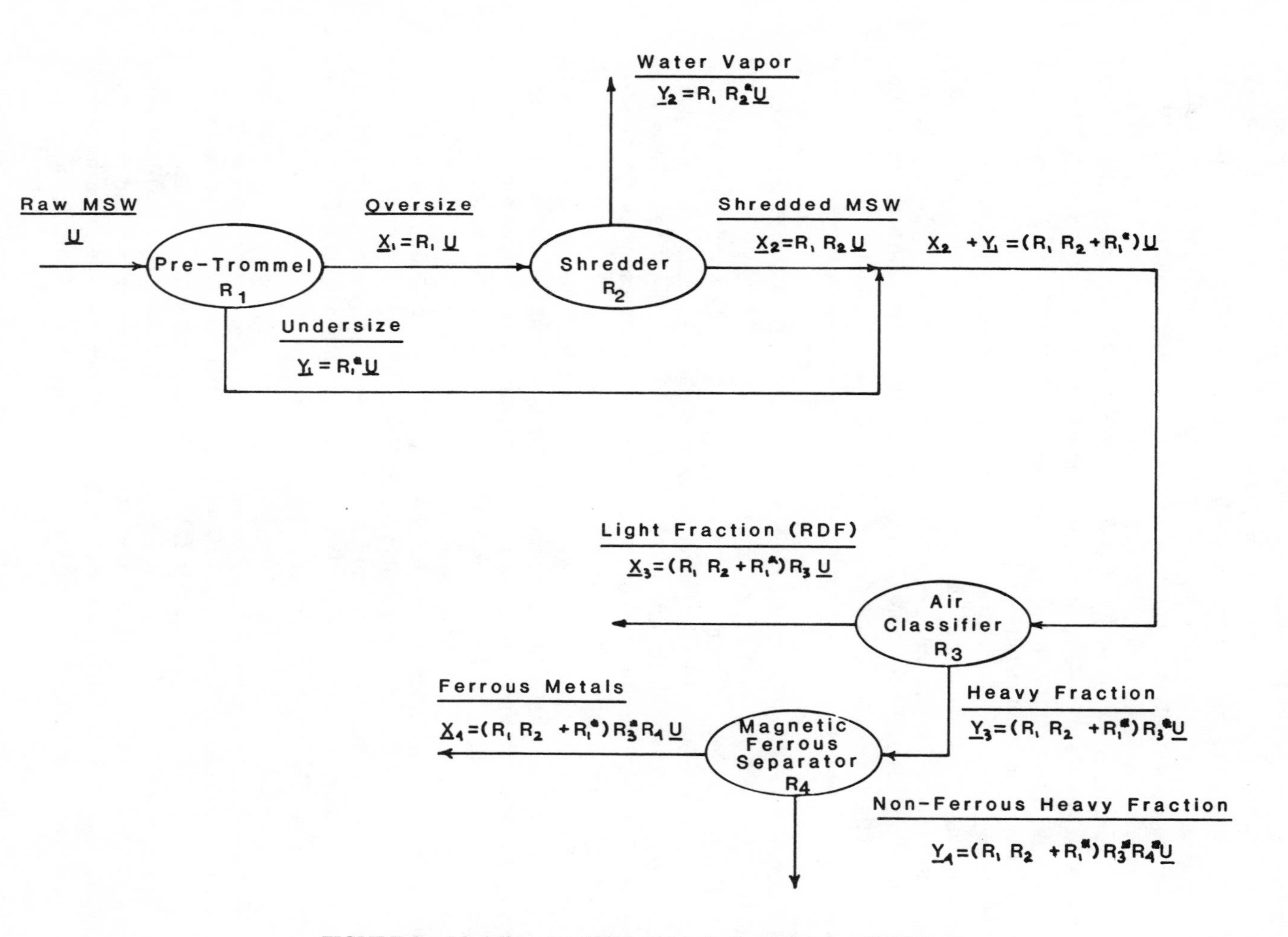

FIGURE 2. Modeling algorithm and block diagram for an RDF facility

Table 3
UNIT PROCESS MASS BALANCE (M_f = MASS FRACTION)

MC[a] (%)	Com-ponent[b]	Raw MSW U Solid M_f	Raw MSW U H_2O M_f	Raw MSW U Solid + H_2O M_f	Shredder RFTF A Solid	Shredder RFTF A H_2O	Shredded MSW X_1 Solid M_f	Shredded MSW X_1 H_2O M_f	Shredded MSW X_1 Solid + H_2O M_f	Air classifier RFTF B Solid	Air classifier RFTF B H_2O	ACLF[c] X_3 Solid M_f	ACLF[c] X_3 H_2O M_f
10	Fe	0.0500	0.0056	0.0556	1.00	0.80	0.0550	0.0045	0.0545	0.10	0.10	0.0050	0.0004
10	Al	0.0100	0.0011	0.0111	1.00	0.80	0.0100	0.0009	0.0109	0.50	0.50	0.0050	0.0004
10	Glass	0.1100	0.0122	0.1222	1.00	0.80	0.1100	0.0098	0.1198	0.60	0.60	0.0660	0.0059
20	Paper	0.3000	0.0750	0.3750	1.00	0.80	0.3000	0.0600	0.3600	0.98	0.98	0.2940	0.0588
10	Plastic	0.0550	0.0056	0.0556	1.00	0.80	0.0500	0.0045	0.0545	0.98	0.98	0.0490	0.0044
20	OIR	0.0300	0.0075	0.0375	1.00	0.80	0.0300	0.0060	0.0360	0.20	0.20	0.0060	0.0012
65	OR	0.1200	0.2230	0.3430	1.00	0.80	0.1200	0.1783	0.2983	0.70	0.70	0.0840	0.1248
Totals		0.6700	0.3300	1.0000			0.6700	0.2640	0.9340			0.5090	0.1959

Component	ACLF solid + H_2O M_f	Cyclone RFTC C Solid	Cyclone RFTC C H_2O	Cyclone ACLF X_5 Solid M_f	Cyclone ACLF X_5 H_2O M_f	Cyclone ACLF X_5 Solid + H_2O M_f	Trommel RFTF D Solid	Trommel RFTF D H_2O	(+) Trommel[d] X_7 Solid M_f	(+) Trommel[d] X_7 H_2O M_f	(+) Trommel[d] X_7 Solid + H_2O M_f
Fe	0.0054	1.00	0.90	0.0050	0.0004	0.0054	0.80	0.80	0.0040	0.0003	0.0043
Al	0.0054	1.00	0.90	0.0050	0.0004	0.0054	0.80	0.80	0.0040	0.0003	0.0043
Glass	0.0719	1.00	0.90	0.0660	0.0053	0.0713	0.20	0.20	0.0132	0.0011	0.0143
Paper	0.3528	1.00	0.90	0.2940	0.0529	0.3469	0.85	0.85	0.2499	0.0450	0.2949
Plastic	0.0534	1.00	0.90	0.0490	0.0040	0.0530	0.90	0.90	0.0441	0.0036	0.0477
OIR	0.0072	1.00	0.90	0.0060	0.0011	0.0071	0.25	0.25	0.0015	0.0003	0.0018
OR	0.2088	1.00	0.90	0.0840	0.1123	0.1963	0.25	0.25	0.0210	0.0281	0.0491
Totals	0.7049			0.5090	0.1764	0.6854			0.3377	0.0787	0.4164

Table 3 (Continued)
UNIT PROCESS MASS BALANCE (M_f MASS FRACTION)

Component	Air classifier RFTC B		ACHF X_7			Trommel RFTF D		(−) Trommel[f] X_8		
	Solid	H_2O	Solid M_f	H_2O M_f	Solid + H_2O M_f	Solid	H_2O	Solid M_f	H_2O M_f	Solid + H_2O M_f
Fe	0.90	0.90	0.0450	0.0040	0.0490	0.20	0.20	0.0010	0.0001	0.0011
Al	0.50	0.50	0.0050	0.0004	0.0054	0.20	0.20	0.0010	0.0001	0.0011
Glass	0.40	0.40	0.0440	0.0039	0.0479	0.80	0.80	0.0528	0.0042	0.0570
Paper	0.02	0.02	0.0060	0.0012	0.0072	0.15	0.15	0.0441	0.0079	0.0520
Plastic	0.02	0.02	0.0010	0.0001	0.0011	0.10	0.10	0.0049	0.0004	0.0053
OIR	0.80	0.80	0.0240	0.0048	0.0288	0.75	0.75	0.0045	0.0008	0.0053
OR	0.30	0.30	0.0360	0.0535	0.0895	0.75	0.75	0.0630	0.0843	0.1473
Totals			0.1610	0.0679	0.2289			0.1713	0.0978	0.2691

[a] Assumed moisture content.
[b] Fe = ferrous;
Al = aluminum;
OIR = other inorganic residue;
OR = other organic residue.
[c] ACLF = air classifier light fraction.
[d] (+) Trommel = trommel oversize.
[e] ACHF = air classifier heavy fraction.
[f] (−) Trommel = trommel undersize.

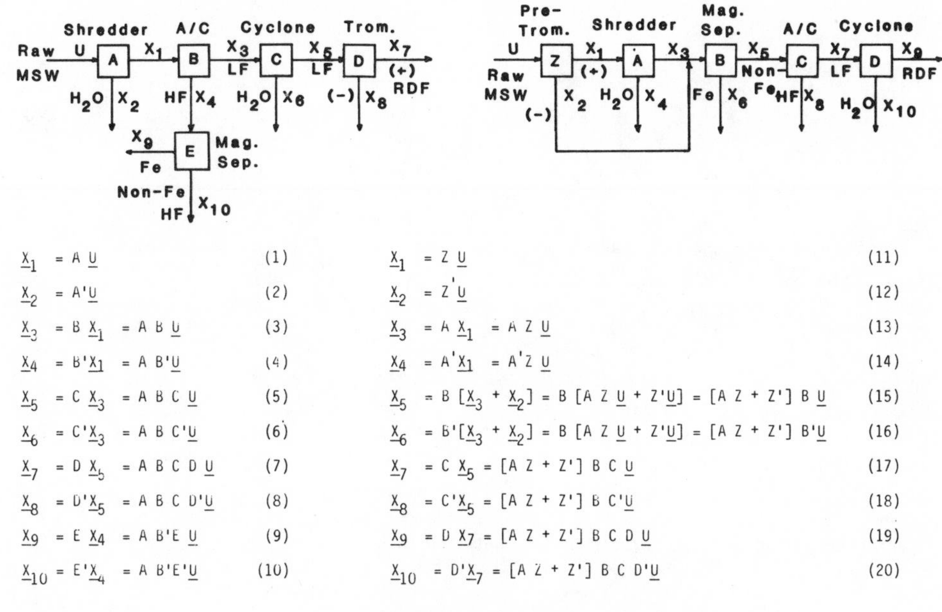

FIGURE 3. Modeling algorithms for two RDF recovery systems—one with and one without pre-trommeling.

Table 4
ASH CONTENTS OF STREAM (M_f = MASS FRACTION)

Assumed component ash content (%)	Raw MSW M_f	Shredded MSW M_f	ACLF M_f	Cyclone ACLF M_f	(+) Trommel M_f	ACHF M_f	(−) Trommel M_f
100	0.05	0.05	0.0050	0.0050	0.0040	0.0450	0.0010
100	0.01	0.01	0.0050	0.0050	0.0040	0.0050	0.0010
100	0.11	0.11	0.0660	0.0660	0.0132	0.0440	0.0528
5	0.15	0.015	0.0147	0.0147	0.01250	0.0003	0.0022
5	0.0025	0.0025	0.0025	0.0025	0.00221	0.00005	0.0002
100	0.03	0.03	0.0060	0.0060	0.0015	0.0240	0.0045
10	0.012	0.012	0.0084	0.0084	0.0021	0.0036	0.0032
Ash weight	0.2295	0.2295	0.1076	0.1076	0.0395	0.12195	0.0649
Dry weight	0.670	0.670	0.5090	0.5090	0.3377	0.1610	0.1713
Percent ash	34.3	34.3	21.1	21.1	11.7	75.7	37.9

Table 5
HEATING VALUES OF WASTE STREAM

	Assumed component heating value[a] Btu/lb	Raw MSW Btu	Shredded MSW Btu	ACLF Btu	Cyclone LF Btu	(+) Trommel Btu	ACHF Btu	(−) Trommel Btu
	0	0	0	0	0	0	0	0
	0	0	0	0	0	0	0	0
	0	0	0	0	0	0	0	0
	8,000	2,400	2,400	2,352	2,352	1,999	48	353
	15,000	750	750	735	735	662	15	74
	0	0	0	0	0	0	0	0
	8,000	960	960	672	672	168	288	504
Total Btu		4,110	4,110	3,759	3,759	2,829	351	931
Wet weight (lb)		1.000	0.934	0.705	0.685	0.4177	0.2289	0.2691
HHV (wet)		4,110	4,400	5,332	5,488	6,773	1,533	3,460
Dry weight (lb)		0.670	0.670	0.509	0.509	0.3377	0.1610	0.1713
HHV (dry)		6,134	6,134	7,385	7,385	8,377	2,180	5,435

[a]Heating value = high heating value = HHV; units = Btu/lb.

Table 6
MOISTURE CONTENTS OF STREAM (M_f = MASS FRACTION)

	Raw MSW M_f	Shredded MSW M_f	ACLF M_f	Cyclone LF M_f	(+) Trommel M_f	ACHF M_f	(−) Trommel M_f
Solids and water	1.000	0.934	0.705	0.685	0.418	0.229	0.269
Water	0.330	0.264	0.196	0.176	0.079	0.068	0.098
Percent	33.0	28.3	27.8	25.7	18.9	29.7	36.3

E. System Comparison Through Modeling

The utility of the recovery factor transfer function approach in modeling processing systems can be well exemplified by using it to make a comparison between a RDF recovery system in which pretrommeling is incorporated, and one in which it is not. Algorithms for both systems are presented in Figure 3. The transfer functions for each unit process are shown sequentially in Table 7, beginning with the assumed composition of the raw waste. It should be noted that the recovery factors listed for the unit processes are based upon values typical of those encountered in practice, and were not derived by way of analytical expressions. The differences between the recovery factors for a given unit process (e.g., air classifier) in a system lacking a pretrommel and those of the same unit process in a system having a pretrommel are mainly the consequence of the changes in particle size distribution brought about by the difference between the two systems in their order of processing. (They are "educated estimates".) In the table, the column under the heading "Composite RFTF" corresponds to the overall transfer function as shown in Figure 3 for the terms X_7 and X_9 (Equations 9 and 19 in the figure) for the system with and without a pretrommel. If as-processed weights are required, the component dry solids and their representative water contents are separated and subsequently added together. The energy and ash yields are calculated respectively with

Table 7
COMPARISON BETWEEN TWO RDF SYSTEMS

SYSTEM WITHOUT PRETROMMEL

	Raw MSW (solids M_f/ H_2O M_f)	Shredder RFTF A (solids/ H_2O)	A/C RFTF B (solids/ H_2O)	Cyclone RFTF C (solids/ H_2O)	Trommel RFTF D (solids/ H_2O)	Composite RFTF ABCD (solids/ H_2O)	RDF X_7 Solids/ M_f	RDF X_7 H_2O M_f	Constituent fuel properties HHV (Btu/ lb)	Constituent fuel properties Ash (%)	Energy[1] yield (Btu/ lb)	Ash yield (lb/ lb)
Fe	0.05/0.0056	1.00/0.80	0.10/0.10	1.00/0.90	0.80/0.80	0.80/0.058	0.004	0.0003	0	1.00	0	0.004
Al	0.01/0.0011	1.00/0.80	0.50/0.50	1.00/0.90	0.80/0.80	0.40/0.288	0.004	0.0003	0	1.00	0	0.004
Glass	0.11/0.0122	1.00/0.80	0.60/0.60	1.00/0.90	0.20/0.20	0.12/0.086	0.013	0.0011	0	1.00	0	0.013
Paper	0.30/0.0750	1.00/0.80	0.98/0.98	1.00/0.90	0.85/0.85	0.83/0.600	0.249	0.0450	8,000	0.05	1,992	0.0125
Plastic	0.05/0.0056	1.00/0.80	0.98/0.98	1.00/0.90	0.90/0.90	0.88/0.635	0.044	0.0036	15,000	0.05	660	0.0022
OIR	0.03/0.0075	1.00/0.80	0.20/0.20	1.00/0.90	0.25/0.25	0.05/0.036	0.0015	0.0031	0	1.00	0	0.0015
OR	0.12/0.2230	1.00/0.80	0.70/0.70	1.00/0.90	0.25/0.25	0.18/0.126	0.0216	0.0281	8,000	0.10	173	0.0022
							0.3371	0.0787			2,825	0.0394

SYSTEM WITH PRETROMMEL

	Raw MSW solids M_f/ H_2O M_f	Pre-trommel RFTF Z (solids/ H_2O)	Shredder RFTF A (solids/ H_2O)	Magnetic seperator RFTF B (solids/ H_2O)	A/C RFTF C (solids/ H_2O)	Cyclone RFTF D (solids/ H_2O)	Composite RFTF (AZ + Z) (BCD) (solids/ H_2O)	RDF X_7 (solids M_f/ H_2O M_f)	Constituent fuel properties HHV (Btu/ lb)	Constituent fuel properties Ash (%)	Energy yield (Btu/ lb)	Ash yield (lb/ lb)
Fe	0.5/0.0056	0.41/0.41	1.00/0.80	0.20/0.20	0.10/0.10	1.0/0.9	0.020/0.018	0.0010/0.0001	0	1.00	0	0.0010
Al	0.01/0.0011	0.37/0.37	1.00/0.80	1.00/1.00	0.50/0.50	1.0/0.9	0.500/0.045	0.0050/0.0001	0	1.00	0	0.0050
Glass	0.11/0.0122	0.01/0.01	1.00/0.80	1.00/1.00	0.02/0.02	1.0/0.9	0.020/0.018	0.0022/0.0002	0	1.00	0	0.0022
Paper	0.30/0.0750	0.69/0.69	1.00/0.80	0.98/0.98	0.98/0.98	1.0/0.9	0.960/0.864	0.2880/0.0648	8,000	0.05	2,304	0.0114
Plastic	0.05/0.0056	0.62/0.62	1.00/0.80	0.98/0.98	0.98/0.98	1.0/0.9	0.960/0.864	0.0480/0.0003	15,000	0.05	720	0.0024
OIR	0.03/0.0075	0.02/0.02	1.00/0.80	1.00/1.00	0.15/0.15	1.0/0.9	0.150/0.135	0.0045/0.0010	0	1.00	0	0.0045
OR	0.12/0.2230	0.43/0.43	1.00/0.80	0.95/0.95	0.40/0.40	1.0/0.9	0.380/0.342	0.0456/0.0763	8,000	0.10	365	0.0046
								0.3943/0.1428			3,389	0.0341

Table 8
SPECIFIC ENERGY REQUIREMENTS OF KEY UNIT PROCESSES

Unit processes	Specific energy (kWh/Mg_i)[a]
Size reduction[b]	$35.5\ X_{90}^{-0.81}$
Air classification	7.0
Magnetic conveyor	0.4
Trommel screen	1.1
Cyclone separator (rotary airlock)	0.3

[a] Mg_i is the mass throughput at unit process "i".
[b] Here the freewheeling contribution has been neglected for purposes of estimation only; the nominal particle size (X_{90}) is expressed in cm.

Table 9
CALCULATION OF ENERGY REQUIREMENTS FOR UNIT PROCESSES IN A RESOURCE RECOVERY SYSTEM[a]

Unit processes	Specific energy[b] (kWh/T_i)	Processed mass fraction (T_i/T MSW)	Energy requirement (kWh/T MSW)
Size reduction[c]	6.6	1.0	6.6
Air classification[d]	7.0	1.0	7.0
Magnetic conveyor	0.4	0.3	0.1
Cyclone separator	0.3	0.7	0.2
Trommel screen	1.1	0.7	0.8
Total			14.7

[a] The system without a pretrommel shown diagramatically in Figure 2.
[b] T_i = tons processed at unit process "i".
[c] X_{90} = 8 cm.
[d] Air classifier mass split: 70% lights, 30% heavies.

the use of constituent heating values and ash contents on a dry solids basis and the constituent mass fractions.

Through the use of values calculated from Table 7, a comparison can be made between the refuse-derived fuels produced by two different processing trains. A summary of such a comparison is made in Table 7. According to the data in Table 7, the ash yield and content and the unit weight of ash per unit of energy are less in the system having pretrommeling than in the one lacking it. On the other hand, the moisture content of the RDF produced with pretrommeling is predicted to be greater than that produced without benefit of pretrommeling.

In summary, the comparisons made in the preceding paragraphs are only a few of those that can be made between systems through the use of the modeling technique described herein.

F. Unit Process Energy Requirements

For purposes of design, some typical values for the specific energy requirements of key unit processes can be established, as is done in Table 8. The specific energy requirements given in the table are to be used in the generic sense, that is, as being indicative of the hypothetical "average" piece of equipment in each unit process category. Within a factor of approximately ±50% the specific energy requirements for air classification, magnetic conveyors, trommel screens, and cyclone separators (rotary airlock) can be considered constant for purposes of estimating system energy requirements. (The blower is considered part of the air classifier system and not as part of the cyclone separation system.) On the other hand, the energy requirements for size reduction are a strong function of the desired particle size of the shredded product as represented by the expression in Table 8.

Using the expression for size reduction and the specific energy values for the other unit processes, it is possible to estimate the energy required for refuse processing. In Table 9 is presented a sample calculation of the energy required by the unit processes in a resource recovery system without a pretrommel. The flow diagram is shown in Figure 2.

REFERENCES

1. **Murphy, J. D.,** Materials to resist the abrasion of pneumatically transported processed refuse, Energy Conservation Through Waste Utilization: Proc. 1978 Natl. Waste Process. Conf., American Society of Mechanical Engineers, New York, 1978, 327.
2. **Diaz, L. F., Glaub, J. C., and Trezek, G. J.,** Laboratory Evaluation of the Impact of Drying and Screening on Refuse-Derived Fuel Quality, prepared for the Electric Power Research Institute, RP 1180-6, April 1981.
3. **Trezek, G. J., Diaz, L. F., Savage, G. M., and White, R.,** Prediction of the Impact of Screening on Refuse Derived Fuel Quality, prepared for Electric Power Research Institute, RP 1180-6, June 1979.
4. **Fiscus, D. E., Gorman, P. G., Schrag, M. P., and Shannon, L. J.,** St. Louis Demonstration Final Report: Refuse Processing Plant Equipment, Facilities and Environmental Evaluations, EPA-600, 2-77-155a, 1977.
5. **Even, J. C., Adams, S. K., Gheresus, P., Joensen, A. W., Hall, J. L., Fiscus, D. E., and Romine, C. A.,** Evaluation of the Ames Solid Waste Recovery System: Part 1—Summary of Environmental Emissions, Equipment, Facilities, and Economic Evaluations, November 1977.
6. **Hopkins, V., Simister, B. W., and Savage, G.,** Comparative Study of Air Classifiers, prepared under EPA Contract No. 68-03-2730, November 1980.
7. **Savage, G. M., Diaz, L. F., and Trezek, G. J.,** Performance Characterization of Air Classifiers in Resource Recovery Processing, Proc. 1980 A.S.M.E. Natl. Waste Process. Conf., May 1980, 339.

Chapter 4

SIZE REDUCTION

I. INTRODUCTION

As used in solid waste management, the term "size reduction" has at least five synonyms, namely, "shredding", "milling", "hammermilling", "comminution", and "grinding". Since only about 25% of the material in refuse is brittle, the term "shredding" probably is about as appropriate as any. As such, it is becoming more widely adopted in references to size-reducing refuse.

Size reduction is an essential step in centralized resource recovery operations. The reduction in size enhances ease of handling, and renders the dimensions of bulky items compatible with those of the processing equipment. Size reduction brings about a degree of uniformity in terms of the maximum particle size of the diverse components of the incoming waste stream. The qualification "degree of uniformity" is used advisedly. While it is true that, in general, the characteristic particle size of all groups of components are reduced by at least an order of magnitude, the size distributions of shredded refuse typically span three to four orders of magnitude. However, some uniformity is a requirement for most mechanical sorting systems. Size reduction is a *sine qua non* in certain processes, as for example, composting, methane fermentation, and hydrolysis of cellulose.

The practice of size reduction in municipal waste management is not solely a resource recovery activity. It is receiving considerable attention as a preparatory step to landfilling. Here the purpose is, of course, to improve the overall quality of landfilling as well as to increase landfill "life." Some go so far as to claim that with the institution of shredding in a fill operation, the need for a soil cover is obviated. The relative advantages continue to be a subject of debate.

II. TYPES OF SIZE-REDUCTION EQUIPMENT

The following types of machines have been designated as shredders by the Waste Management Equipment Manufacturers Institute: crushers, cage disintegrators, shears, shredders, cutters and clippers, hammermills, and grinders. Of this collection, the hammermill is the type most widely used in refuse processing.

A. Hammermills

The hammermill is a type of impact crusher in which a combination of tensile, compressive, and shear forces is applied to the throughput material by striking particles of the material while they are in suspension, or by hurling the particles at a high speed against stationary surfaces. In some sense, the greater part of the size reduction of the refuse is accomplished by brute force.

Hammermills can be divided into two broad groups on the basis of orientation of the rotor, namely horizontal and vertical. Both types can have either rigid or swing hammers. The horizontal swing hammer, whose principal parts are the rotor, hammers, grates, frame, and flywheel, is the type most commonly used. Its rotor and flywheel are mounted through bearings to the frame. The bottom portion of the frame also holds grates. A diagrammatic sketch of a hammermill is shown in Figure 1. The hammers may have one of a number of configurations, depending upon the particular manufacturer. When the entire rotor-hammer assembly is rotating (usually at 1000 to 1500 rpm) the

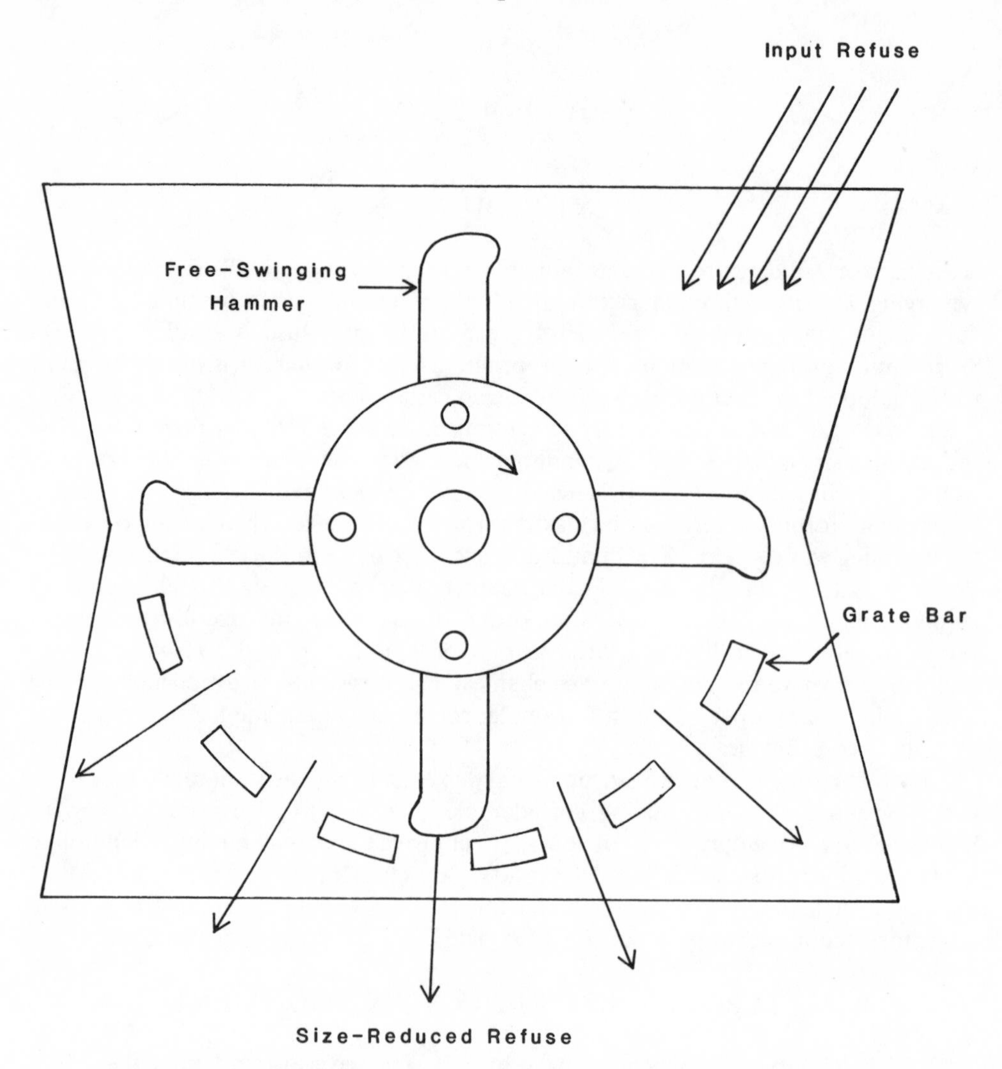

FIGURE 1. Internal arrangement of hammers and grates in a horizontal hammermill.

hammers fly perpendicularly to the rotor. Upon impact with the objects being size reduced, each hammer is free to move in a 180° arc within the plates of the rotor. Materials to be size reduced enter the machine through an in-feed chute and interact with the hammers and each other until at least one dimension of the objects has reached a size small enough for the particle to fall through the grates at the bottom of the machine.

The residence time of the material in the mill as well as the size distribution of the product are largely determined by grate spacing. Other factors that affect product size distribution are feedrate, moisture content, and mill speed.

In a vertical hammermill, the rotor is placed in vertical position, as is indicated in Figure 2. The input, assisted by gravity, drops parallel to the shaft axis. A variation of the vertical principle is embodied in the ring grinder (e.g., Eidal). The design of a ring grinder differs from that of other hammermills in that size reduction is accomplished by a set of gear-like teeth installed in a rotor which fits in a stationary ribbed housing. Material introduced into the upper part ("throat") of the machine drops into a set of breaker bars by means of which large objects are torn apart. The material then enters

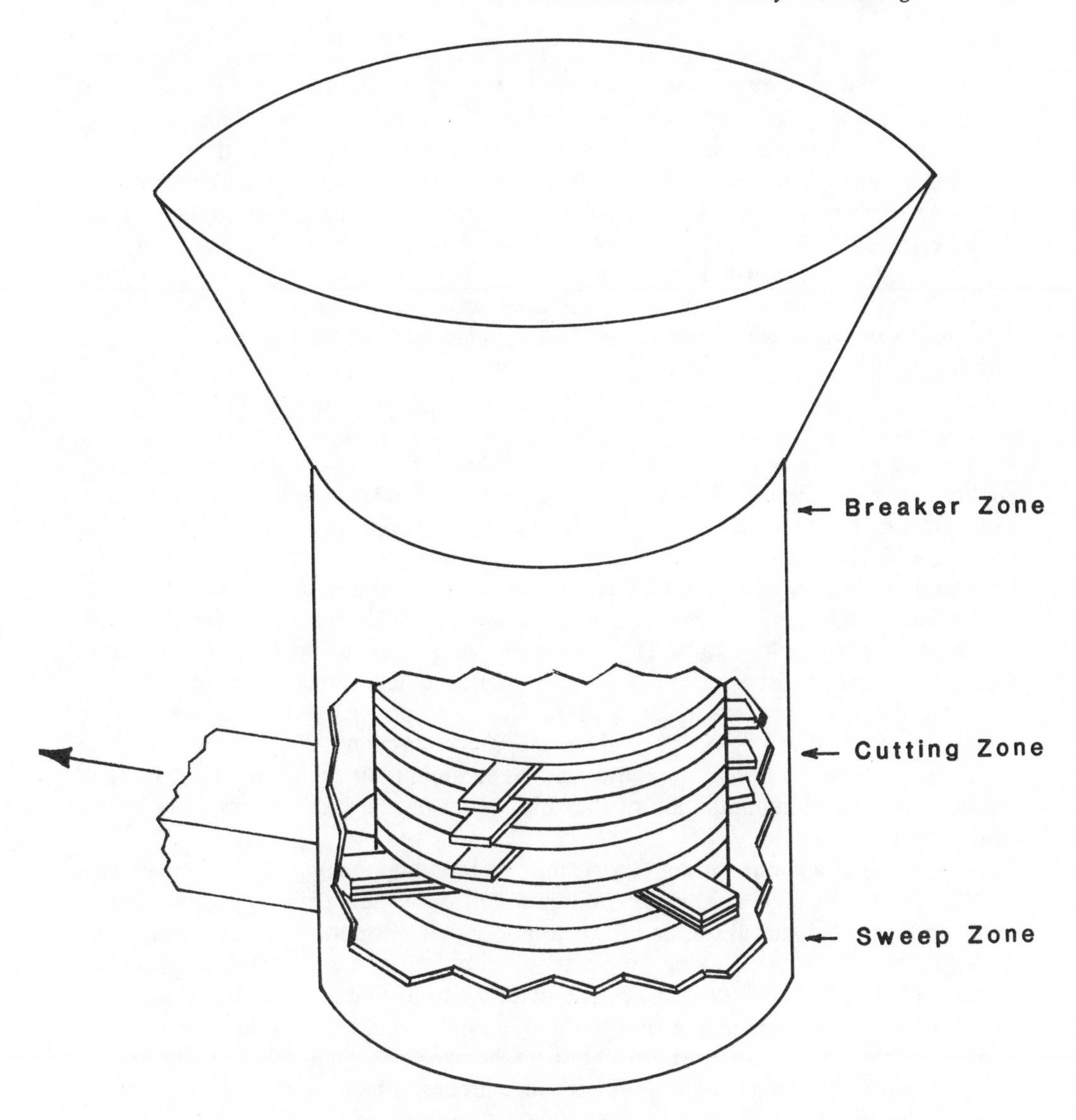

FIGURE 2. Vertical hammermill.

the space between the teeth and the housing ribs where size reduction is accomplished as a result of induced shear and a self-comminution interaction between the various particles passing through it. Comminuted material drops out at the bottom.

III. PROCESS PRINCIPLES

A. Disruptive Forces

1. Basic

The three types of basic forces involved in size reducing municipal solid wastes are shear, tension, and compression. All three forces play an important part in shredding municipal refuse, because it is a heterogeneous collection of materials characterized by a wide range of physical properties in terms of brittleness or lack thereof, ductility, and hardness.

The application of tension to a body usually is accompanied by compression and shear forces which result from gripping forces and the internal transfer of energy. The closer the opposing tension forces are together, the lower will be the breaking energy. Tension forces develop at points separated by relatively small distances in a hammermill in which a hammer impacts a mass and breaks off a section. In compression, the fracture of a body is caused by internal tension and shear forces that are a result of the compression force. As with tension forces, the energy used is the force applied multiplied by the distance traveled. When exerted upon a piece of material, shear forces usually are accompanied by both tension and compression forces. Compression forces are applied to a body in offset planes to produce a shearing action. The shearing of a flat material may be accomplished with a minimum of work because of the relatively short distance traveled by the force.

2. Analysis and Interaction

Although the size reduction of refuse is a technology that is finding an increasing application in municipal solid waste management, the tendency in developing designs and modeling the interaction between the forces involved continues to be a reliance upon empiricism, rather than upon fundamental principles. As a consequence, the existing sizeable literature on size reduction and its principles is based almost entirely on size reducing homogeneous brittle materials, whereas refuse is a conglomeration of brittle and nonbrittle materials. The reason for the apparent reluctance to take into account the characteristics of refuse is the pronounced heterogeneity of the material, especially that of the physical make-up of the diverse objects in it. Although empiricism has its place in design and modeling, advanced processing cannot be optimally designed solely on an empirical basis. The reason is that comminution events are involved that require control and manipulation of the subsequent product size distributions along with process scaling.

To be general, a comminution theory must be based upon a variety of factors. For example, dominant mechanisms of comminution in any grinding machine influence the final product size distribution, and in turn, are dependent upon machine design and particle characteristics. The particle size distribution is the summation of many single-event distributions occurring as a result of the particular grinding action peculiar to the mill. A common failing of the theories and so-called laws discussed in the literature is that they fail to give information on the size-throughput relations of a mill or of an optimum operating condition. A breakage process effects changes throughout all sizes of a complete size distribution. A single parameter (e.g., surface area) cannot of itself describe changes of such complexity.[1] In the course of time, attempts were made to fill the gaps in the existing knowledge, although the attention continued to be centered on homogeneous collections of brittle materials. Progress made along those lines is described in a review by Snow.[2] More recently, studies dealing with mixtures and especially heterogeneous materials are beginning to be reported.[3-4]

Pertinent details of concepts described in the literature are presented herein to elucidate possible extensions of principles and knowledge of size reduction of brittle materials to the comminution of nonbrittle heterogeneous materials.

B. Initial Energy Comminution Relations

Early theories on the energy aspects of size reduction centered on attempts to arrive at an index for a given material that could be related to a standard range of size reduction, and that could be used to compute the energy required to produce any other range of size reduction. This index, c, which depended physically on material properties and mill operating variables, is mathematically described by the relation

$$\frac{dE}{dx} = -\frac{c}{x^n} \tag{1}$$

where E is the net energy required per unit weight in a given process of comminution, x is a factor related to size, and n is an exponent. An inherent problem with this index stems from the difficulty of expressing size in a mathematical form, since in any actual process, the material being comminuted has a considerable range of size on entering and leaving the process. Moreover, it is impossible to denote this size by a single index because of variations in the shapes of the particles.

Three relations frequently quoted for comminution derived from Equation 1 are those attributed to Rittinger, Kick, and Bond. They are obtained as follows: When n = 2, the integration of Equation 1 becomes an expression of Rittinger's law:

$$\Delta E = c_R \left(\frac{1}{x_2} - \frac{1}{x_1}\right) \tag{2}$$

When n = 1, the integration expresses Kick's law:

$$\Delta E = c_K \log \left(\frac{x_1}{x_2}\right) \tag{3}$$

Bond's third theory of comminution is obtained by substituting n = 1.5 in Equation 1, integrating, and rearranging it to arrive at

$$\Delta E = w_i \frac{R^{1/2} - 1}{R^{1/2}} \left(\frac{100}{x_2}\right)^{1/2} \tag{4}$$

where w_i is a work index and R is the size reduction ratio x_1/x_2. Obviously, the energy requirement in Bond's theory must be between the two predicted by the two earlier laws.

The three theories differ with respect to definitions of x_1 and x_2. According to Rittinger's theory, the energy consumed in comminution is proportional to the increase in surface area produced (surface theory). Therefore, x_1 and x_2 must be defined as the particle sizes that have a surface area equal to the total surface area of a unit mass of feed or product divided by the total number of particles contained in that mass. Kick's theory, namely that analogous changes of configuration of geometrically similar bodies vary as the volumes of the bodies (volume theory), implies that x_1 and x_2 are particle sizes that have their mass equal to the mean mass of the particles in the feed and product. Bond defines his particle sizes as the apertures through which pass 80% by weight of the feed or product.

Rittinger's theory is applicable primarily to the crushing of relatively small particles, whereas Kick's theory is suited to large particles. However, the evidence indicates that the numerical value of n depends not only on the material being crushed, but also on the design of the crusher. Bond made an allowance for the latter variation through the work index, w_i. Bond's determination of the work index for a given material involved arriving at an arithmetic mean value of energy per ton as based on data accumulated from observations made of several types of crushers and grinding equipment, each having its own characteristic efficiency. Despite his data being so widely scattered as to be seemingly unsuitable for taking the arithmetic mean, the numerical value 1.5 for n made Bond's approach more workable than either Kick's or Rittinger's theories when it came to correlating data related to a number of crushing devices. Bond's theory should be considered useful, principally to codify a mass of industrial experience so that an interpolation and limited extrapolation can be made for known materials and equipment.

Charles,[5] Holmes,[6] and Schumann[7] have shown indirectly that energy-size relations can have as a common base the simple relationship

$$\Delta E = Ck^{-m} \tag{5}$$

where C is a machine constant, k is the size modulus (theoretical maximum size from size distribution curve), and m is a constant.

In summary the energy-size relationship as described in the pre-1970 literature does not provide a complete theoretical analysis of the comminution process. Although in some instances the approach proposed by the workers of that period may provide a basis for a partial correlation of experimental data, it is inadequate for meaningful process simulation.

C. Size Distribution Relations

Among the empirical size-distribution equations formulated to describe the physical nature of the comminuted product are the equations by Gaudin[8] and Schumann,[9] and another by Rosin and Rammler.[10] The Gaudin-Schumann equation

$$Y(x) = \left(\frac{x}{k}\right)^{\alpha} \tag{6}$$

relates the fraction of undersize Y(x) to relative size x/k. In this relationship, Y(x) is the cumulative fraction by weight finer than the stated size x, and k is a quantity called the size modulus (theoretical maximum size), which together with the slope parameter, α, characterize the product. The slope parameter usually is close to unity.

The Rosin-Rammler equation relates the same relative fraction of undersize to the relative size modulus by an equation of a different form, namely,

$$Y(x) = 1 - \exp\left[-\left(\frac{x}{x_o}\right)^{n}\right] \tag{7}$$

where n is some numerical constant, and x_o is a characteristic particle size, i.e., 63.2% cumulative passing.

A theoretical size-distribution equation for single fracture as proposed by Gaudin and Meloy[11] is the following:

$$Y(x) = 1 - \left(1 - \frac{x}{\bar{x}_o}\right)^{r} \tag{8}$$

In Equation 8, $\bar{x}_o$ is a characteristic dimension of the feed particle before fracture, and r is the size ratio, which is a measure of the number of breaks in the particle. Bergstrom[12] modified the Gaudin-Meloy equation empirically by adding another parameter and arrived at the following equation:

$$Y(x) = \left[1 - \left(1 - \frac{x}{x_o}\right)^{r}\right]^{q} \tag{9}$$

Equation 9 embodies a combination of the fine and coarse-size curve-fitting capacities of the Gaudin-Schumann and Gaudin-Meloy equations.

Other size distribution equations have been proposed with additional parameters, as further generalizations of the above relations, ostensibly for greater curve-fitting flexibility. However, space does not allow a further discussion on the subject.

1. Matrix Analysis of Breakage

Because refuse is a heterogeneous mixture, there is little possibility that all of the types of breaking that occur in its size reduction can be described by a single breakage "law". Indeed, a major deficiency with the preceding theoretical size distribution relations is that experimental results indicate that the feed cannot be characterized by any of them (the relations). A possible solution to the problem is to use the matrix method, especially to simulate the breakage process so as to arrive at predictions of feed and discharge size distributions.

With the matrix method, the size distribution based upon determination by sieve analysis is replaced by a vector. This is done by considering the range of size of the collection of particles to be subdivided into n internals, with the largest size being x and the smallest as x_{n+1}. The ith size fraction is bounded by x_i above and x_{i+1} below and $x_i = rx_{i+1}$ [(i = 1, 2, 3, . . . , n − 1), with r (> 1)] being the geometric sieve ratio. Where it is necessary to specify the size of particles in a particular size fraction by a single number, the geometric mean is taken. When the size scale is logarithmic, the geometric progression of sizes appears as equally spaced points on the abscissa.

P_i refers to the cumulative mass fraction of product material below x_i, and F_i is the cumulative mass fraction of feed material below size x_i where i = 1, 2, 3, . . . , n. Then, the feed and product size distributions can be described by n × 1 column vectors $\underline{f}$ and $\underline{p}$ with elements f_i and p_i, where f_i and p_i are, respectively, mass fractions of feed and product materials that fall between x_i and x_{i+1}, thus:

$$\underline{f} = \begin{matrix} f_1 \\ f_2 \\ \bullet \\ \bullet \\ \bullet \\ f_n \end{matrix} \quad ; \quad \underline{p} = \begin{matrix} p_1 \\ p_2 \\ \bullet \\ \bullet \\ \bullet \\ p_n \end{matrix} \tag{10}$$

with

$$f_i = F_i - F_{i+1} \tag{11}$$

$$p_i = P_i - P_{i+1} \tag{12}$$

The percentage or fraction of material under the smallest size, the undersize, is described by

$$f_{n+1} = F_{n+1} = 1 - \underline{e}'\underline{f} \tag{13}$$

$$p_{n+1} = P_{n+1} = 1 - \underline{e}'\underline{p} \tag{14}$$

where $\underline{e}'$ is the transposition of the n × 1 column vector $\underline{e}$ with unity elements. The transformation of the feed to the product is thus expressed mathematically by matrix algebra:

$$\underline{p} = \underline{\underline{X}}\,\underline{f} \tag{15}$$

where $\underline{\underline{X}}$ expresses the breakage process.

The sequence of rows and columns in the matrix, $\underline{\underline{X}}$, which describes a transformation of a size distribution, bears definite relations to particle size. If i and j are referred to the ith row and jth column of $\underline{\underline{X}}$, with elements X_{ij}, then X_{ij} will denote the proportion of particles that initially were in size range j and afterwards were in product size range i.

For size reduction, $\underline{\underline{X}}$ is lower triangular, and a knowledge of the n elements of $\underline{p}$ and the n elements of $\underline{f}$ is not sufficient to arrive at a unique solution from the n equations of the $\frac{n}{2}$ (n + 1) unknowns X_{ij} ($i \geq j$).

Another approach to the empirical evaluation of the matrix of an open-circuit grinding process is essentially described in the equation

$$\underline{p}_j = \underline{\underline{X}}\ \underline{e}_j \qquad (j = 1, 2, 3, \ldots, n) \tag{16}$$

where $\underline{\underline{X}}$ is unknown and $\underline{e}_j$ is the unity vector with all elements zero except the jth element (= 1). The experimental determination of $\underline{p}_j$ gives the jth column of X. If particle interaction does not occur, and if it is possible to use graded feeds ($\underline{e}_j$) down to the smallest size x_n, the experimental evaluation of $\underline{\underline{X}}$ should satisfactorily predict the output size distribution ($\underline{p}$) obtained on feeding ungraded material ($\underline{f}$).

2. *Breakage Process Models*

A matrix method similar to that developed by Callcott[13] for size reducing coal in a swing hammermill may be adapted to the analysis of the size reduction of refuse. Generally, when refuse is size reduced in a swing hammermill, under condition of low through put most of the input tends to pass through the mill without being exposed to repeated fracture. (However, it could be that operational conditions greatly different from those considered would lead to a more extensive recirculation, and hence repeated exposure.) In a "continuous" type of operation, the residence time of material in the machine is brief. Consequently, the swing hammermill can be reasonably considered as a "once-through grinder" for the simulation of the refuse shredding process, as was done by Callcott for the comminution of coal. Of four models based on such a matrix method,[14] the two that most approach reality are summarized as follows.

a. π—Breakage Process

The π breakage is one in which the selection function S(Y) equals π at all sizes Y, in that a proportion, π, of the particles, independent of particle size, is broken according to the breakage function B(x, y). In terms of the matrix, $\underline{\underline{S}} = \pi\underline{\underline{I}}$, and the equation describing the process is

$$\underline{p} = [\pi\underline{\underline{B}} + (1 - \pi\underline{\underline{I}})\ \underline{f}] \tag{17}$$

The parameter π completely describes the process if $\underline{\underline{B}}$ is assumed to be known, and it is a useful measure of the breakage accomplished in the process. Techniques for back-calculating the value of π from known feed and product distributions are described by Broadbent and Callcott.[15] However, they state that the simple simultaneous solution of the elements of Equation 17 usually should not be attempted, because of the errors that might attend measurements. They outline a least-squares technique as a better approach.

b. Repeated Breakage Cycles

When the feed size distribution differs considerably from that of the product, the breakage process can be assessed in terms of a series of cycles of mild breakage. The description may be written as

$$p = \prod_{h=1}^{N} (\underline{\underline{BS}}_h + \underline{\underline{I}} - \underline{\underline{S}}_h)\ \underline{f} \tag{18}$$

where h equals 1, 2, . . . , N refers to the hth cycle in which the probability of breakage is $\underline{\underline{S}}_h$. If only $\underline{p}$, $\underline{f}$, and $\underline{\underline{B}}$ are given, it becomes impossible to evaluate each $\underline{\underline{S}}_h$ and N.

However, the following sequential model may be used to simplify the analysis. If it is assumed that the probability of breakage $\underline{\underline{S}}_n$ is the same for all the cycles, Equation 18 becomes

$$\underline{p} = (\underline{\underline{B}}\,\underline{\underline{S}} + \underline{\underline{I}} - \underline{\underline{S}})^N \underline{f} \tag{19}$$

where $\underline{\underline{S}}$ is the selection matrix common to all the N cycles. If

$$D = (\underline{\underline{B}}\,\underline{\underline{S}} + \underline{\underline{I}} - \underline{\underline{S}}) \tag{20}$$

then

$$\underline{p} = \underline{\underline{D}}^N \underline{f} \tag{21}$$

The entire size-reduction process that takes place in the mill is treated as a sequence of operations in which the product from the jth cycle becomes the feed for the (j + 1)th cycle. To ensure that the undersize from any cycle of the process is correctly described, $\underline{p}$ and $\underline{f}$ are (n + 1)-column vectors, with the last elements, p_{n+1} and f_{n+1}, describing the undersize for the product and feed, respectively. $\underline{\underline{D}}$ is written as a (n + 1) square matrix, with the (n + 1)th row describing the proportions of the various size ranges that end up as undersize.

Callcott[13] considered the hammermill to include N obstructions or grinding zones. The product from the jth zone in passing through the (j + l)th zone is said to have undergone a single cycle of breakage. In the present model, the number of cycles, N, is not identified with the grinding zones. Instead, it is introduced arbitrarily as a means of repeatedly applying the single matrix $\underline{\underline{D}}$ to the feed vector $\underline{f}$ to obtain $\underline{p}$ when the top sizes of the feed and product distributions differ considerably. If x_1 is the largest size of the initial feed and x_m the largest size of the final product (where $x_i = rx_{i+1}$, i = 1, 2, . . . , m, . . . , n − l), then N is estimated from

$$N = (m - 1) \tag{22}$$

c. Discussion

A good model meets certain criteria, among which are the following: (1) it includes the more important features of a process; (2) as far as is possible, it is mathematically simple; (3) it relies upon a minimum of assumptions; and (4) it is useful with respect to accurate prediction and theoretical speculation. In terms of the four criteria, the π-breakage model excels the repeated-breakage-cycle model, although both lead to very good results regarding the products of the primary, secondary, and tertiary grinding process. Nevertheless, of the two models, the π-breakage model is the superior because it leads to results that are more realistic with respect to all grinding conditions.

IV. REGULATING HAMMERMILL PERFORMANCE

Shredder performance can be characterized on the basis of the parameters particle size distribution of the product, specific energy consumption (kWh/Mg), and machine wear. These characteristics are manifestations of the physics of the comminution mechanisms. The important operational factors that influence the three dependent variables are a set of five independent variables, namely, feed size distribution, the flow of material through the device, moisture content of the materials, grate or extraction spacing, and the relative velocities of the size-reduction apparatus in the machine (e.g., tip velocity of the rotating hammers).

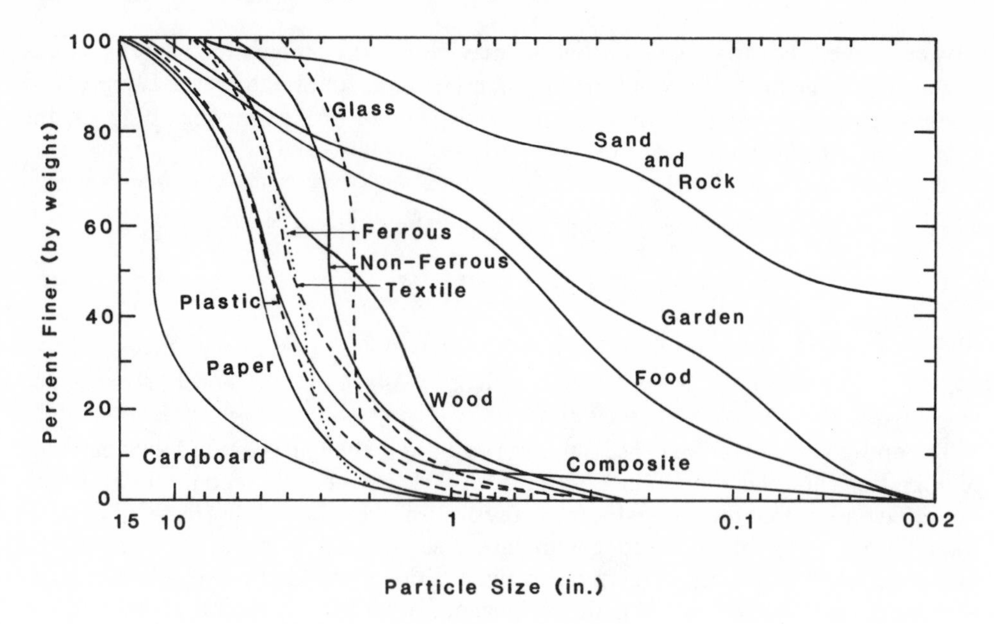

FIGURE 3. Size distribution of components of raw MSW.

A. Particle Size Distribution

The waste stream contains a multitude of different sized objects which collectively cover a range of sizes that spans an order of magnitude. Thus, the material is said to have a certain size distribution. With respect to raw refuse, a range of 90% cumulative passing between 30 and 62.5 cm is usual. The size distribution of a typical municipal refuse and of its components is indicated by the curves in Figure 3. In evaluating shredder performance, the term "characteristic particle size" refers to the size of the screen openings through which 63.2% of the particles can be passed. For example, a waste having characteristic particle size of 40 mm implies that 63% of its particles have a particle size of 40 mm or less. Among the principal factors that affect size distribution and its characterization are theoretical representation, number of passes through the mill, rotational velocity of the hammer tips, moisture content of the input material, grate spacings, and the condition of the hammers (degree of wear).

1. Theoretical Representation

The ability to represent product size analytically makes it feasible to develop a semiempirical theory in terms of breakage and selection functions with which to predict product size distribution for any feed. Although the theory thus developed can be extended to embrace a study of the interrelation and effects of the collection of comminution variables, confining it to the examination of the size distribution itself provides a useful insight into the behavior of the other variables.

Experience[16-21] indicates that product distributions can be represented by a Rosin-Rammler relation of the type:

$$y(x) = 1 - \exp\left[-(x/X_o)^n\right] \tag{23}$$

where y(x) is the cumulative percent passing screen size x. The exponent n ("n index") is essentially the slope of the line $\ln[1/(1 - y)]$ vs. x on log-log coordinates. The value of X_o is a size that corresponds to 63.2% cumulative passing, or alternatively, the value of x where $\ln[1/(1 - y)] = 1.0$. The individual values of n obtained under all grinding conditions observed thus far fall within the limits ($0.5 \leq n \leq 1.3$) of the values characteristic in the grinding of brittle materials in hammermills.[16-20] Lower n values usually are associated with a rather scattered distribution over a wide size range. On the other hand, high n values and a uniform particle structure over a narrow size range seem to go together. The Rosin-Rammler distribution also physically indicates that at a given value of X_o, the cumulative percent passing a size x (i.e., [y (x)]) decreases as the value of n increases.

2. *Number of Passes*

The product size distribution shifts towards the finer sizes with successive cycles of grinding. For example, in observations made by Savage et al.[16] and by Trezek and Savage[17, 18] of a hammermill in which the rotor revolved at 1200 rpm, it was found that the characteristic particle size (X_o) ranged from 1.5 to 2.5 cm in primary grinding (i.e., the first pass), and the n value was 0.99. In the secondary grinding (second pass), the X_o dropped to 0.825 cm, and the n to 0.87. In the third grinding, X_o became 0.575 cm, and n, 0.75.

3. *Rotational Velocity*

The effect of rotational velocity of the hammer tip (tip speed) is interpreted here in terms of rpm of the rotor. This interpretation is valid because the results on which it is based were obtained with one and the same machine. In other words, the hammer radius was constant throughout. The machine was designed such that it could be operated at 1200, 790, and 555 rpm. The n indexes at each rpm were 0.81, 0.97, and 0.91, respectively, and the corresponding X_o values, 1.95, 2.7, and 2.2 cm, respectively. While these numerical values may vary from machine to machine because of differences in hammer dimensions and machine design, the relationship remains comparable.

Numbers obtained in studies[16-20] do indicate a production of a coarser product at lower velocities. A likely reason is that as the rpm is increased, so is the number of impacts per unit of time. Of course the force of the impact also increases with increase in velocity. The values for X_o and n more or less "plateau" at the two lower velocities, perhaps because velocity ceases to participate in a predominant manner in the overall description of the size-reduction process.

4. *Moisture Content*

Because of diametrically opposed findings reported in the literature, a firm statement regarding the effect of moisture content on particle size distribution cannot be made at this time, excepting to state that it does have a definite effect. The nature of this effect is where the disagreement comes in. In one study[17] a definite increase in coarseness of particle size was observed to accompany increase in moisture content within the observed range, namely 37 to 63%. On the other hand, a second group of researchers in another locale[21] observed the reverse, namely, a smaller average particle size and a more uniform size distribution with increase in moisture (i.e., 32 to 48%). Reasons in favor of the second group's findings are (1) fibrous constituents tend to become weak upon being wettened; and (2) frictional resistance of refuse is lessened at the higher moisture contents, and thus greater shearing forces can be developed. Factors that may negate their findings stem from certain physical characteristics that are increasingly aggravated as shredded refuse becomes more moist. These factors result in a tight clumping of the particles to such an extent as to make an accurate size distribution measurement well nigh impossible to obtain, especially when the moisture content exceeds 60%.

5. Condition of Hammers

It is likely that hammer condition may be a more significant factor than is moisture content in determining particle size distribution. The size distribution of particles becomes increasingly coarse as the hammers lose their cutting edges, i.e., become worn. The change is greatest immediately after new or resurfaced hammers are installed. A reason for the increased coarseness is the progressive widening of the space between the hammer tips and the housing that results from the gradual wearing down of the hammers. Moreover, worn hammers do not shred as effectively as new hammers.

The wear effect on size distribution has been described in terms of exponential decay with time.[21] In the description, the relation for percent passing (y) takes the form:

$$y = b_o + b_1 \exp(-b_2 t) \tag{24}$$

where t is the cumulative tons of refuse shredded since the installation of new hammers.

B. Specific Energy Consumption (Power Requirement)

Energy consumption is a key element in the practical and economic feasibility of any shredding process. It is the quantity of gross minus freewheeling energy divided by throughput rate. When considering power requirements, a distinction should be made between gross power, net power, and freewheeling power. The gross power refers to the total power required, and the net power, to the gross power minus the freewheeling power. The freewheeling (idle) power represents the power required when no refuse is in the machine. Typically, about 5 to 10% of the full load rating is freewheeling power, and therefore from 85 to 90% is available for size reduction. Key factors that affect energy consumption by a shredder are characteristic particle size, feed rate, moisture content, machine design (includes) hammer placement, and state-of-repair of the machine.

1. Characteristic Particle Size

Within a range of n values from 0.7 to 1.3, the specific energy consumption (kWh/Mg) is a function of characteristic size, in that it is at its greatest when the particle size (X_o) is small, and least when particle size is large. Thus with coarser particles, the amount of energy to produce a given percent in X_o is less than that with finer particles. For example, at one installation it was found that to obtain a 50% reduction in characteristic particle size (i.e., from 15 cm down to 7.5 cm) only about 2.42 kWh/Mg were required, whereas to make a similar reduction at an X_o of 1.5 cm (i.e., from 1.5 cm down to 0.75 cm) required a power input of 7.16 kWh/Mg.

With information available at the time of this writing, it is possible to develop relationships with which predictions can be made on an as-received basis regarding specific energy as a function of characteristic size (X_o) and of the nominal (X_{90}) product size. (Nominal size refers to 90% cumulative passing.)

In the development of a functional relationship between E_o and the size parameters, the following two equations were formulated:

$$E_o = 17.91\ X_o^{-0.90} \tag{25}$$

$$E_o = 35.5\ X_{90}^{-0.81} \tag{26}$$

in which E_o is expressed in kWh/Mg (as received), and X_o and X_{90} in cm.[17,18] The corresponding correlation coefficient for Equation 23 is 0.87, and for Equation 26, 0.89. Inasmuch as the equations are based on data relating to throughputs that range from 15- to 82-Mg refuse per hr, they can be of assistance in arriving at estimates of net power requirements in general. However, it should be emphasized that the estimates will be

Table 1
RANGE OF COEFFICIENT VALUES SHOWING EFFECT OF FLOW RATE AND MOISTURE CONTENT IN FULL-SCALE OPERATIONS

Shredder	Correlation coefficient	Number of data points	Equation
Appleton east	0.85	11	$P_N = 6.33\ \dot{m}^{0.94}(1 - MC)^{1.54}$
Ames primary	0.74	10	$P_N = .62\ \dot{m}^{1.20}(1 - MC)^{-7.04}$
Ames secondary	0.93	10	$P_N = .28\ \dot{m}^{1.92}(1 - MC)^{-2.84}$
Cockeysville no. 1, forward	0.92	6	$P_N = 47.04\ \dot{m}^{0.82}(1 - MC)^{4.79}$
Cockeysville no. 1, reverse	0.96	6	$P_N = 6.04\ \dot{m}^{1.18}(1 - MC)^{2.01}$
Cockeysville no. 1, combined	0.91	12	$P_N = 21.85\ \dot{m}^{0.90}(1 - MC)^{2.25}$
Great Falls 20 TPH	0.28	12	$P_N = 39.97\ \dot{m}^{0.27}(1 - MC)^{-0.63}$
Tinton Falls	0.85	12	$P_N = 4.05\ \dot{m}^{0.94}(1 - MC)^{1.66}$
Odessa, forward	0.91	12	$P_N = 0.14\ \dot{m}^{0.84}(1 - MC)^{-12.64}$

Modified from Savage, G. M. and Shiflett, G. R., Processing Equipment for Resource Recovery Systems, III. Field Test Evaluation for Shredders, final report by Cal Recovery Systems, Inc., for U.S. EPA, Cincinnati, Ohio, 1980.

generalities because a number of other variables also influence power requirements. An estimate can be made of the net power required to produce a specific particle size by multiplying the specific energy (E_o) by the anticipated flow rate. The product thereof divided by 0.9 is a rough estimate of the required gross power.

2. Flow Rate

The relation of flow rate to power required can be represented by the equation

$$P_N = a\ \dot{m}^b (1 - MC)^c \qquad (27)$$

where P_N = net power (kWh); $\dot{m}$, the flow rate of refuse through the shredder (Mg/hr); and MC, the fractional moisture content of the refuse.[22] A positive value for the exponents b and c indicates that power increases with flow rate, but decreases with moisture content. The coefficient a determines the magnitude of the power requirements, and is influenced by machine design characteristics such as size of grate openings, internal geometry, and the geometry and number of hammers. The range of coefficient values that could be encountered in full-scale operations is indicated by the data in Table 1. (The data were obtained during the course of an investigation of the performance of shredders in the field.[22]) A perusal of the nine equations listed in the table shows that in all of them the calculated value of the exponent b is positive, and with two exceptions approach unity. From this fact, it can be concluded that at least with respect to the operations referred to in the table, the net power requirements were in direct proportion to flow rate. The first exception from the trend towards unity was in the form of an upsurge in the b value characteristic of one of the operations (Ames secondary). The upsurge could have been a function of the extremely worn condition of the hammers that was noted at the time the observations were made. The second exception (Great Falls plant) manifested itself in the opposite direction, namely a sharp decline. The observers offer no explanation for the latter departure, and state that the power drawn by the machine proved to be almost independent of feed rate.

Experience at the time of this writing indicates that the power demand drops as the moisture content of the throughput wastes increases, although there is some indication that energy consumption begins to increase again when the moisture exceeds 50%. The

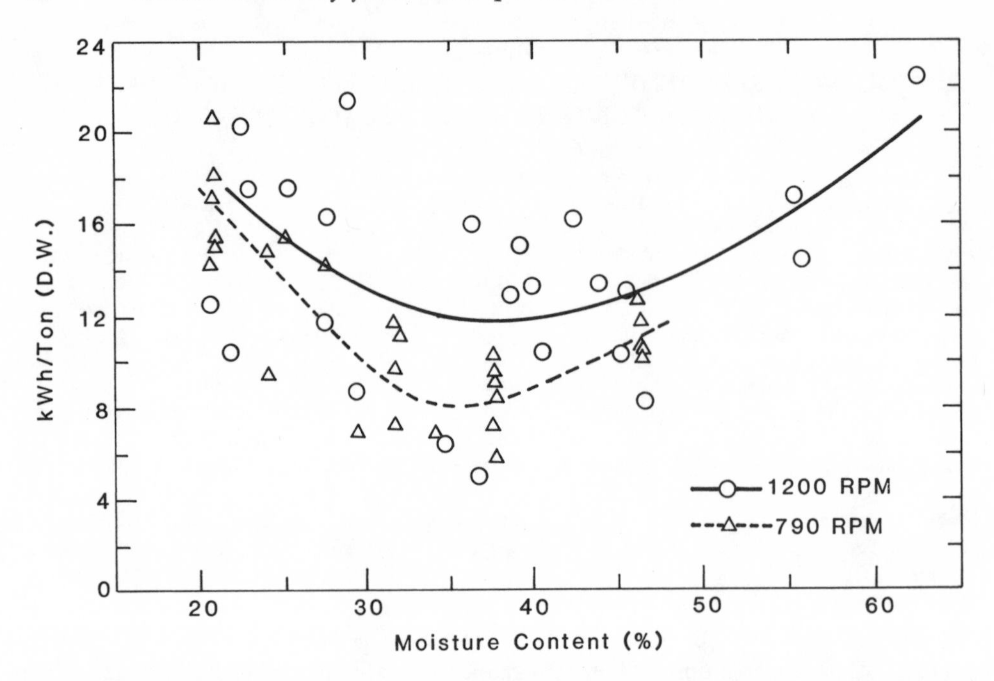

FIGURE 4. Effect of moisture on specific energy consumption. (Adapted from Ruf, J. A., Particle Size Spectrum and Compressibility of Raw and Shredded Municipal Solid Waste, Ph.D. dissertation, University of Florida, 1974.)

relationship is illustrated by the curves in Figure 4. Incidentally, the figure also shows the lowering effect had by decline in rotor speed on energy requirements.

There is some discrepancy between the relations indicated by the sign of the exponent c with equations in Table 1 and the data reported in the literature. In the equations, the effect of moisture content is indicated by the sign for the exponent c. If the sign is positive, then the indication is that energy requirements decline with increase in moisture content. Conversely, if the sign is negative, then an increase in power requirement with drop in moisture content is indicated. The discrepancy is in the fact that the exponent is negative in four of the equations. Thus far, the resolution of the discrepancy awaits further research.

3. Machine Design

As one would expect, the design of the machine, overall and individual components, has a definite bearing on energy requirements. Obviously, the better the design, the more efficient the machine is in terms of energy usage. The effect is demonstrated by the existence of the wide range in the numerical values of coefficient a in the equations in Table 1, a range which extends from 0.14 to 47.04.

Shredder "holdup" is a measurement that holds promise as a useful parameter for determining machine characteristics upon energy consumption and throughput capacity.[19,22] "Holdup" is a term that refers to the amount of material within the shredder at any instant in time. The greater the amount of holdup, the greater is the power requirement. The relation of holdup to power input is shown by the slope of the curve in Figure 5. Available experimental and field-scale data can be made to fit the following two curves:

$$P_N = 1.60\ H^{0.96} \tag{28}$$

$$\dot{m}_w = 0.51\ H^{1.04} \tag{29}$$

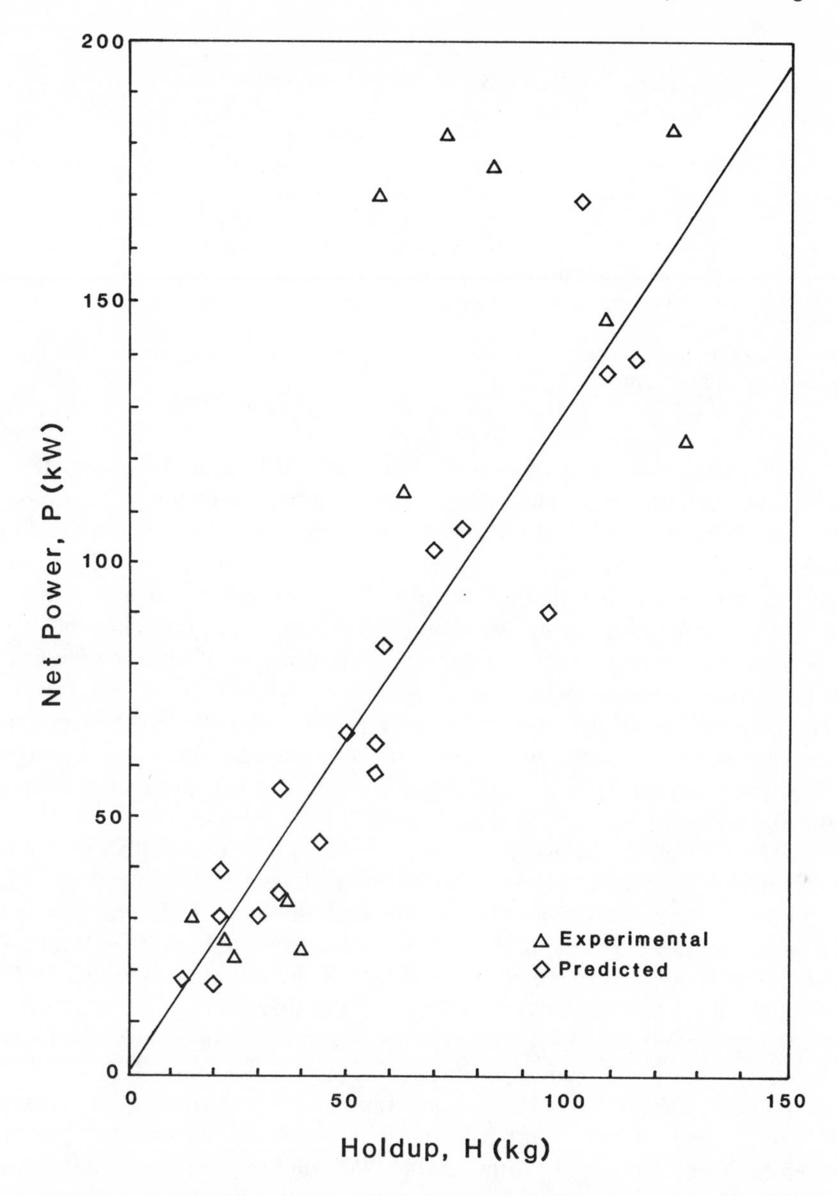

FIGURE 5. Relation of "holdup" to power requirement. (From Shiflett, G. R., A Model for the Swing Hammermill Size Reduction Refuse, D. Eng. dissertation, University of California, Berkeley, 1978. With permission.)

The placement of hammers in a machine can have an effect on reducing the specific energy consumption. For example, replacing a 54-hammer arrangement in a certain model of the vertical Tollemache grinder by a 34-hammer configuration resulted in a substantial reduction in power needs.

C. Machine Wear-and-Tear

An exceedingly important feature of municipal solid waste (i.e., refuse) rarely, if ever, named in tables on the composition of refuse is its extreme abrasiveness. The magnitude of abrasiveness is evidenced by the fact that hammers tipped with the hardest of feasibly applied alloys are worn to an unusable condition after a throughput of only about 1000 Mg, which is little more than a couple of days throughput in the average

Table 2
PRINCIPLE GROUPS OF HARDFACING ALLOYS

Group	Description
1	2—12% Alloy content (soft and ductile)
2	12—25% alloy content (Cr and Mo are major alloying elements)
3	25—50% alloy content (Cr is the major alloying element)
4	Co- and Ni-based alloys
5	Carbide-based alloys (extremely hard and brittle deposits)

Note: The facings increase in hardness in the order listed; thus, group 1 is ductile, whereas group 5 is extremely hard and brittle.

processing installation. Observed erosion rates of 0.013 to 0.106 kg/Mg of refuse processed have been reported. The attrition resulting from abrasion by refuse is to be distinguished from the mechanical and structural failures resulting from overloading the machine with a bulky item or an oversized piece of tramp metal. The latter can be avoided by good management practice, whereas attrition erosion is an unavoidable fact that arises from the prevailing abrasiveness of refuse. This latter intimates the key part had by attrition in determining the economics of shredding, and hence the feasibility of a resource recovery operation.

The two components of the shredder that are subjected to the most extensive and intensive wear-and-tear are the hammers and the grate bars. This is to be expected because the two components come into direct contact with the throughput wastes and constitute the working face of the machine.

A number of factors, all relatively obvious, exert an influence on attrition. Although mostly obvious, few have been precisely defined in terms of design, operation, and maintenance parameters. Among the factors that might be mentioned is type of material being processed. For example, hammer wear seems to be more extensive when the refuse has a large percentage of paper. Wear is less intensive when the moisture content is high. Wear becomes less pronounced if the characteristic particle size of the product is large. Correspondingly, at a given output particle size, the rate of attrition diminishes in proportion to reduction in particle size of the input. Rotor speed also seems to have an effect in that wear is less extensive at lower speeds. For example, a 31% decline in material lost from hard-tipped hammers was observed to have taken place when the rpm was reduced from 1200 to 790 rpm. The drop with bare hammers was 43%.

Probably the most important factor in determining rate of attrition is the hardness of the metal used in making the hammers. If the hammers are tipped in alloy, it is the hardness of the facing. Groups of alloys available for hardfacing are identified and described in Table 2. Factors that determine which of the groups is best suited for facing include cost, ease of application, compatibility with the base hammer material, and abrasion and impact resistance. A problem is that costs rise in proportion to the hardness of alloy. Experience has shown that the useful "life span" of a hammer tipped with a subgroup 2c alloy is from 30% (at 790 rpm) to 57% longer than that of an untipped hammer. Conceivably, the prolongation might be from 100 to 150% if groups 3, 4, and 5 were used as facing. The limiting factor at the higher degrees of hardness is a tendency for chipping of the welds to occur at high impact loadings.

Savage et al. report that in their investigations[16,22] wear was at its greatest with hammers of manganese or carbon steel (hardness 14 R_c or less) that were not hardfaced. For example, in one operation, it was observed that with the hardness at 11 R_c, the hammer wear in the form of weight loss per hammer was 3.33×10^{-3} kg/hammer-Mg of

Table 3
HAMMER MAINTENANCE COSTS (1978 dollars)

Cost items	Unit cost per ton (U.S. dollars)	Annual Cost (U.S. dollars)	Percent of cost
Case I: Buildup Once Per Week			
Labor			
Remove and install	0.052	4,900	11.9
Welder	0.269	25,400	61.8
Materials			
Hardfacing	0.077	7,300	17.8
Hammers (2 sets)	0.037	3,500	8.5
Totals	0.435	41,100	100.0
Case II: Daily Buildup			
Labor			
Welder	0.322	30,400	73.6
Materials			
Hardfacing	0.097	9,100	22.0
Hammers (1 set)	0.019	1,800	4.4
Totals	0.438	41,300	100.0
Case III: Wear and Scrap			
Labor			
Remove and install	0.013	1,200	2.6
Materials			
Hammers (1 set)	0.483	45,600	97.4
Salvage credit	(0.036)	(3,400)	—
Totals	0.487	43,400	100.0

refuse processed, whereas at 51 R_c, it amounted to 1.48×10^{-3} kg/hammer-Mg. Apparently, the optimum range for hardness is within the range of 48 R_c to 56 R_c. The facing that would be most suitable to a specific shredding operation would have to be determined experimentally by testing a group of different hardfacing alloys, the collective hardness of which would span the optimum range. In making the test, all alloys to be tried should be tested simultaneously.

1. Relation of Maintenance to Costs

A corollary to its effect on rate of wear of hammers is the relation of the refacing to cost of hammer maintenance. However, in this case cost is not solely a function of that of the alloy, it also is strongly influenced by operational procedures. The bearing of operational procedures is illustrated by the following example, which is based on actual testing. Three procedures were followed. The first involved building up (resurfacing, retipping) the hammers once each 5 days of operation. In the second, the hammers were retipped each day (during the second shift). A wear-and-scrap program constituted the third procedure. In a wear-and-scrap program, the hammers are allowed to wear down to a point at which the incoming refuse cannot be effectively shredded. The pertinent data are presented in Table 3.

While the analysis given in the example may seem clear-cut, certain important factors are ignored by restricting the analysis to costs alone. They are (1) variations between operations with respect to other procedures; (2) availability of suitable equipment for changing and welding hammers; (3) properly trained maintenance personnel; and (4) the fact that badly worn hammers, inevitable with the wear-and-scrap method, affect particle size, energy requirements, and throughput. There is the strong possibility that the two means most effective in minimizing costs of hammer maintenance are to see to it that maintenance personnel are properly trained, and that suitable equipment is available for maintenance.

V. EVALUATING PERFORMANCE IN FIELD

The utility of performance evaluation needs no elaboration in this book. It is a part of the machine selection process, a part of maintenance, and of research into the principles of design. Performance is evaluated on many bases, the weight of which individually depends upon the interests of the one making or ordering the evaluation. As far as a shredder is concerned, power consumption, throughput capacity, characteristics of the product, and stamina of the machine would obviously be the elements of major importance.

A. Monitoring Power

Power input and consumption can be monitored by an instrument having the following features, or by one that can be adapted to provide them. The equipment used in the work reported in references 16-19 and 22 was a Scientific Columbus Model DL34-2K5A2-AY-6070® watt/watt-hr transducer. The transducer should be integrated with a chart recorder and a digital dividing circuit.

The transducer itself should be adaptable to single phase, to three phase-three wire, or to three phase-four wire systems. Both the voltage and the current in each leg are monitored continuously in multiphase systems as a means of automatically correcting for the power factor. The two output signals provided by the transducer are (1) an analog current signal that is directly proportional to power (kW); and (2) a digital signal that is directly proportional to energy consumption. The latter signal is sent to the dividing circuit through which pulses are counted, and an event marker on a recorder is activated after a predetermined number of pulses have occurred. The divider may be set manually to count from 1 through 9999 input pulses before triggering the event marker.

Appropriate current and voltage step-down transformers must be provided to reconcile the electrical characteristics of the transducer and its ancillary instruments with those of the plant being observed. Reliable chronometers capable of precise measurements down to split seconds should be used in time studies. The type used in the studies in references 16-19 and 22 was a Chronos Model 3-ST® digital chronometer.

Measurement of time is an important step in determining duration of sampling period and the speeds of conveyor belts. Length of sampling time is a key factor in estimating throughput rate (e.g., kg/min). In the absence of a specifically designed device, a method for determining belt speed is to measure the time required for the belt to travel a given distance. The measured distance is divided by the elapsed time. A more direct procedure is to use an instrument designed for measuring conveyor belt speeds, such as the TAK-ETTE Model 1707® digital rpm gauge sold by Power Instruments, Inc.

Screens (sieves) conventionally used in determining particle sizes are suitable for use in arriving at determinations of particle size distributions. The screens should collectively include the following openings (square mesh, expressed in cm): 0.13, 0.27, 0.51, 0.95, 1.59, 2.54, 5.08, 10.16, and 20.32. SWECO® screens are typical.

Table 4
RELATION OF GRATE SPACING TO MASS OF SAMPLE

Grate spacing (cm)	Sample size (kg)
20	10
15	7.5
10	5
5	2.5

B. Methodology

Flow rate samples and power level data should be collected under a variety of operating conditions. A precise coincidence between the collection of the flow rate sample and the monitoring of the power can be ensured by accurately making the following two measurements (1) distance from the center line of the shredder to the center line of the discharge conveyor from which the flow rate sample was gathered; (2) the speed of the discharge conveyor itself. With these two measurements, it is possible to calculate the time required for the sample to move from the shredder to the point at which the sample is collected. The mass flow rate is the weight of sample removed from a given length of belt times the speed of the belt. The power required to produce a specified particle size is equal to the product of specific energy (E_o) times the throughput.

The importance of obtaining a representative sample cannot be overstressed when it comes to determining size distribution. A useful safeguard is to work with a large sample. A useful indicator of appropriate size is the grate spacing of the shredder. In general, the wider the spacing, the larger should be the sample. An indication of the preferred amounts is given by the data in Table 4. The size distribution of the particles making up the sample is determined by screening. Before screening, however, the moisture content of the sample should be determined. This is done by weighing and drying the material to constant weight in a drying room maintained at 22°C and 65% relative humidity. The "dried" material is then reweighed, and the moisture content is estimated on the basis of the difference. It should be noted that the preceding procedure differs from that which is conventionally followed.

Manual or mechanical screening or both may be used in making the size distribution. A standardized procedure for determining the sieve size of refuse-derived fuel particles is being developed by ASTM. Until the ASTM procedure is developed and formally accepted, the following procedure will suffice: the dried refuse is placed on the largest of the manually held screens (i.e., 20.3 cm) and is shaken until all of the undersize particles have fallen through. The material retained on the screen ("oversize") is weighed, and the undersized material is processed on the screen with the 10.16-cm openings, and shaking and appropriate weighing is applied, as was done with the larger screen. The process is repeated down to the screen with the 2.56-cm openings. The material that passed through the 2.54-cm openings is processed through a SWECO® screen system (e.g., SWECO model LC18533333 Vibro Energy Rotary Screen®). The SWECO® system has a series of wire mesh screens through which particles are separated on the basis of retention at screen openings 2.56 cm, 1.59 cm, 0.95 cm, 0.51 cm, 0.27 cm, and 0.13 cm.

1. Hammer Wear

Hammers must be cleaned and weighed immediately before and after the given amount of solid waste is shredded. Degree of wear is expressed as weight of material lost per unit weight of refuse shredded. When testing alloys, base hammer material and hardfacing alloys should be tested simultaneously in each run. This measure is an insurance that the conditions of exposure for all materials were identical.

When making comparisons in terms of hammer wear, it is necessary to take into account the extent of the size reduction accomplished. It is a general rule that wear increases with increase in spread between the particle size distribution of the input material and that of the product. This would be expected because the finer the size, the longer is the exposure time of the average particle to the hammers.

Differences in the particle size distribution of feed material and that of the shredded product can be accounted for through the use of the parameter, "degree of size reduction" (Z_o). The term Z_o is defined as

$$Z_o = (F_o - X_o)/F_o \tag{30}$$

where F_o is characteristic feed size, and X_o, product size.

Values for extent of size reduction extend from zero (no size reduction) to a maximum limit of 1.0 corresponding to a product size of zero, and therefore to an infinite amount of size reduction. Of course, the latter limit is beyond achievement in practice.

The application of the principles discussed in this section to the making of comparisons between single and multistage size reduction (or simply between two plants) in terms of gross (P_G) and net (P_N) power consumption, mass flow rates of refuse ($\dot{m}$), and characteristic product size (X_o) is illustrated by the following example. Numbers used in the example are based upon actual operational data.[22] The flow rate in plant 1 is 47.35 Mg/hr and in plant 2, 19.6 Mg/hr. The equivalent product sizes of the product from the single-stage plant (plant 1) and from the two-stage plant (plant 2) served as the criterion required for making the comparison. (The X_o value was 1.6 cm in both cases.) Since a specified Moisture Content (MC) must be given, the average value (20.5%) of the refuse processed at the two plants is used in the pertinent equations.

On the basis of an MC of 20.5%, Equation 27 can be reduced to

$$P_N = 0.62\ \dot{m}^{1.20}(0.795)^{-7.04} \tag{31}$$

and

$$P_N = 0.28\ \dot{m}^{1.92}(0.795)^{-2.84} \tag{32}$$

for the plant 2 primary and secondary shredders, respectively, where $\dot{m}$ is 21 Mg/hr, a throughput at which an average characteristic product size of 1.6 cm is produced. For the shredder in plant 1, the equation for net power as calculated on the basis of metric tonnes is

$$P_N = 21.85\ \dot{m}^{0.90}(0.795)^{2.25} \tag{33}$$

where $\dot{m}$ is 57.5 Mg/hr, the throughput at which the characteristic product size is the required 1.6 cm.

Through the solution of Equations 31, 32, and 33, the total net power required to produce the required product size (1.6 cm) at plant 2 is 352.6 kW, and at plant 1, 504.7 kW. Taking the throughput material on a wet weight basis, the specific energy ($E_o = P_N/\dot{m}$) for plant 2 becomes 16.3 kWh/Mg, and for plant 1, 9.7 kWh/Mg. Therefore, the specific energy consumed in producing a product size of 1.6 cm at plant 2 is 168% of the energy used at plant 1 to accomplish an equivalent reduction.

However, since net energy represents only a part of the energy consumed, to be

Table 5
SPECIFICATIONS OF HAMMERMILLS FOR COMPARISON OF PERFORMANCE

Specification	Mill 1	Mill 2
Mill length	2.59 m	2.74 m
Motor		
Size	373 kW	746 kW
Type	Both identical	
Grate		
Spacing	24.1 × 36.5-cm openings	20.3 × 35.6-cm openings
Total opening	3.79 m^2	3.54 m^2
Number of hammers	14	24

complete the comparison must include the gross power consumption. The gross energy requirement by the plant 2 two-stage shredding is 456.2 kW, that is, 352.6 kW net power plus 58.4 kW freewheeling energy by the primary shredder and 45.2 kW by the secondary shredder. The gross requirement at plant 1 is 577.4 kW (504.7 kW net power plus 72.7 kW freewheeling power). Therefore, the gross energy consumed in producing a 1.6-cm product is 23.28 kWh/Mg at plant 2 (throughput 19.6 Mg/hr). In the actual case illustrated here, the energy consumption per megagram of refuse in a two-stage size reduction is almost twice that used in a single-stage process.

An important *caveat* to keep in mind when considering the preceding example is that the data are characteristic of two individual installations, and are *not* averages or medians of several observations. Therefore, conclusions based on the data are restricted to the two operations in question and cannot be validly extrapolated to become generalizations. In fact, Savage et al.,[16] in their studies, observed that under certain conditions of grate opening and particle size, the total energy consumed in multiple-stage shredding could be less than that in single-stage shredding.

A second example, again using numbers obtained in actual operations, is given to demonstrate the application of the principles involved in making comparisons between performance and various machine designs. Design specifications of the two horizontal hammermills selected for comparison are listed in Table 5.

As in the first example, characteristic product size is made the basis of comparison. In an examination of available data on the performance of the two mills, no statistically significant relation could be found between product characteristic size and flow rate with respect to mill 2. On the other hand, an inverse relationship between the two can be detected in the data pertinent to mill 1. The latter relationship can best be represented as

$$X_o = -0.034\,\dot{m} + 6.03 \qquad (34)$$

The absence of the correlation between characteristic size and flow rate in the data from mill 1 can be compensated as follows: Equation 34 and the data it describes are plotted as in Figure 6. To compare the data from mill 1 with those from mill 2, an auxiliary axis for the mill-2 flow rates is drawn below the flow rate axis for the mill-1 data, as is shown in Figure 6. Then the mill-2 data are plotted such that the averaged characteristic size of its product fall[5] on the line that describes the mill-1 data. Since the average flow rate and characteristic size of the mill-2 product were 49.5 Mg/hr and 2.1 cm, respectively on the basis of the slope of the curve and the axis for the mill-1 data, the flow rate through mill 1 to produce a product having an average characteristic size of 2.1 cm, would be 114.3 Mg/hr.

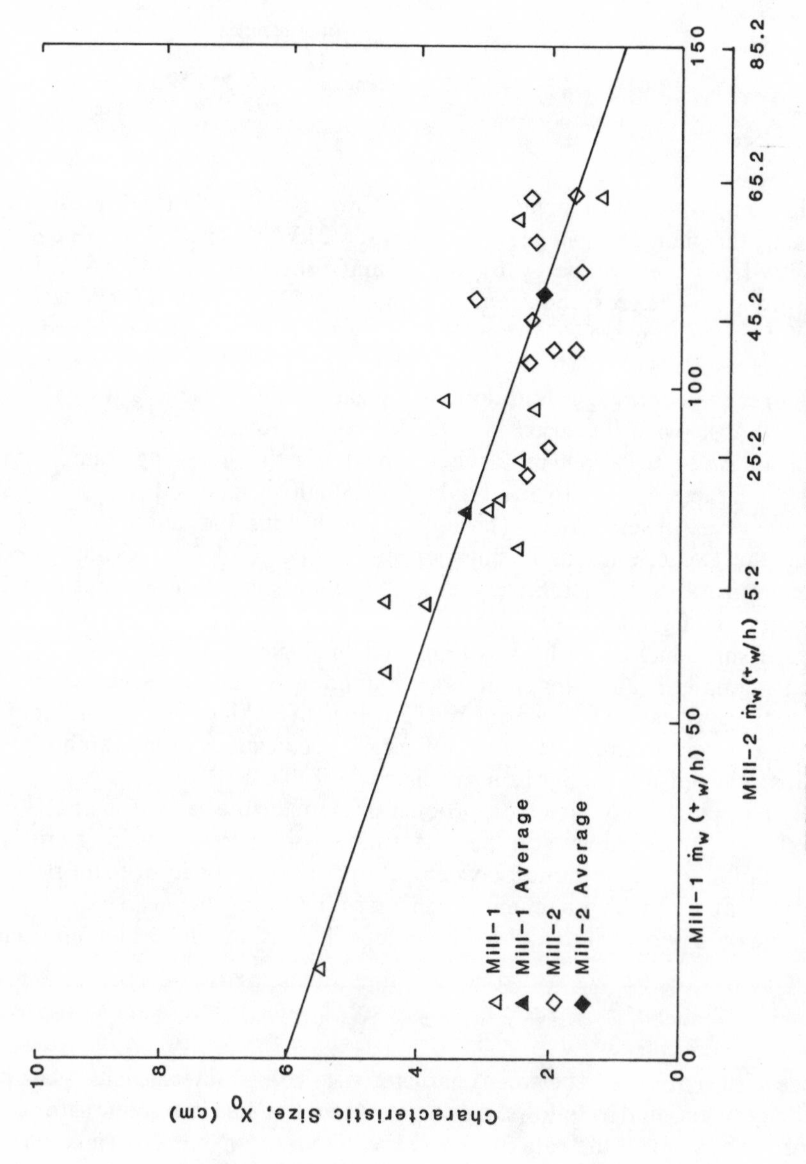

FIGURE 6. Observed and derived flow rate-size relationships.

If the moisture content of the raw refuse fed both mills is 20%, then the net power consumed by mill 1 to reach a characteristic particle size of 2.1 cm would be 124.9 kW, and by mill 2, 441.7 kW (see Equation 27). Consequently, the specific energy required by mill 1 is 1.1 kWh/Mg, and by mill 2, 8.9 kWh/Mg. In the absence of a substantial difference between the characteristics of the input refuse to the two mills, one would have to conclude that this large difference between the two mills in efficiency of energy utilization is a function of differences in design features.

VI. SELECTING A SHREDDER

A. General Criteria

The criteria to be followed in the selection of a shredder are several in number and are of a type applicable in the selection of any piece of equipment , shredder, or otherwise. A criterion that immediately comes to mind is the extent to which the performance of the machine matches that attributed to it by the vendor in terms of throughput rate, power consumption, maintenance, and product characteristics. A second criterion follows from the first—the product produced by the shredder must be suited to its (the products) intended use. For example, the particle size distribution must be within a specified range of dimensions. Meeting the product characteristics criterion is not enough. The shredder must be capable of doing so with a reasonable uniformity and continuity over a specified period of time. In other words, the machine must have a high degree of reliability. Reliability relates to maintenance, and hence another criterion is that the shredder should require an amount of maintenance at least not in excess of that which is normal for such a machine. Moreover, the maintenance should be of a type that is easily applied. An extremely important criterion is an efficient use of energy in terms of energy per unit mass of output. The final criterion to be mentioned here concerns economics, namely *total* or *overall* cost per unit mass of output. *Total* and *overall* are emphasized because preparation and postshredding costs, if substantial, can negate any gain from a seemingly low cost per unit of output from the shredder.

The principles described in the above section, "Evaluating Performance," can be advantageously followed in applying the criteria mentioned in the preceding paragraph. Moreover, they would have a genuine utility when making the visits to existing installations that constitute an essential part of a well-planned selection process.

B. Type of Machine

A problem that complicates the task of selecting an appropriate machine is that from a design standpoint, the relations between physical characteristics of a hammermill and its performance parameters (e.g., specific energy, hammer wear) is only dimly understood at present. This area of uncertainty especially pertains to the relative advantages or disadvantage of the horizontal as compared to vertical hammermill.

One certainty stands out, however. It is the fact that of all the solid wastes, municipal refuse is a leading contender in terms of difficulty of grinding and wear-and-tear on the machine. Therefore, sturdiness of construction should be an important factor in the selection.

At the level of knowledge existing at the time of this writing, no hard and fast statements can be made regarding specific design features and power consumption, although some tentative associations can be made, as were done in the above section, "Evaluating Performance." Thus, chances are that a lowering of the rpm of the rotor is accompanied by a drop in power consumption. But, the penalty for so-doing is a coarser-sized product.

To promote proper maintenance, the machine and machine set-up should include mechanical or hydraulic pin pullers for pin removal and installation, an overhead crane for hammer removal and installation, and an hydraulically activated system for opening the shredder, and all working parts should be readily accessible and replaceable.

1. Vertical vs. Horizontal Hammermills

From a design standpoint, two observations can be made concerning the relative merits of vertical and horizontal hammermills (1) a relative abundance of data is available on the performance of horizontal hammermills; (2) the possession of a grate cage by the horizontal hammermill provides a means of developing analytical relationships between grate opening size, percent open area, throughput, and product size. Grates (or a screen) in the bottom of a horizontal hammermill provide a convenient means of controlling the size of the shredded product. The control is readily attained by varying the size of the spacing between the grate bars or the size of the openings in the screen, as the case may be. On the other hand, to adjust the size of the particles being discharged from a vertical mill, typically it is necessary either to vary the number, shape, and clearance of the hammers, or to change the size of the discharge opening, or to rely upon a combination of all four. The nonspecific nature of such control measures renders their modeling difficult. Furthermore, since control of product size is more difficult to attain, the required experimentation is greater in extent. The difficulty of particle size control using a vertical hammermill in an RDF pelletizing line has been reported by tests conducted by the National Center for Resource Recovery.[23]

2. Machine Components

a. Motor

To be realistic, an estimate of the size of the motor needed to power a particular shredder at a specified throughput rate must be based on gross power requirement, i.e., the net power required to size-reduce the waste *plus* the freewheeling power. The freewheeling power is the "idling" power, i.e., the power consumed by the machine when no refuse is being milled. Typically, idling power accounts for 5 to 10% of the total energy consumed by a machine.

The net power required to produce a product of a particular particle size can be estimated by multiplying the specific energy (E_o) by the anticipated flow rate ($\dot{m}$). The product divided by 0.9 constitutes a rough approximation of the necessary gross motor power. It should be noted that the estimate does not include the safety factor needed to provide for chance surges in the rate of input to the machine or for the presence of objects that are difficult to grind. In the U.S. shredder motors typically were being operated at 13 to 73% of the full-load motor rating at the time of this writing. The average was about 36%.

b. Hammers

Important considerations in the selection of hammers are number of hammers per machine, metallic composition of the basic hammer and of its facing (cutting edge), configuration of the individual hammers, placement, and ease of access, removal, and replacement.

An austenitic manganese steel is a good base hammer material. Its toughness and abrasive resistance makes the steel a suitable choice for withstanding the combined effects of abrasion resulting from glass, dirt, and paper, and of impacts upon concrete and metallic objects in the refuse. Obviously, if a hammer consisting of such a steel is to be hard faced, then the alloy used for the facing must be compatible with the steel. Basically, the groups of alloys listed in Table 2 can meet this requirement. As stated earlier, the harder the alloy, the more resistant it is to abrasion. On the other hand, because of brittleness, the harder alloys tend to be chipped off the hammers by impact with hard objects.

C. Establishment of Required Specific Equipment Specifications

1. Basic Principles

Although shredding has long been a basic unit process in resource recovery, it is only recently that energy consumption and machine wear have been recognized as being significantly related to the composition and particle size of the shredded product. Recovery facilities equipped with shredders that are either too large or too small are not at all uncommon. Frequently, a shredder and motor in a facility may be incompatible with each other. An example of incompatibility is a shredder that has an internal capacity (i.e., "cavity") insufficiently large to permit the desired degree of size reduction at a given throughput, and yet is provided with an oversized motor.

In order to proceed with a rational design or selection of a shredder, three basic criteria must be specified. The first criterion refers to the characteristics of the input material, i.e., type and physical and mechanical properties such as moisture content, bulk density, composition, and particle size distribution. The second criterion refers to the desired size distribution and configuration of the shredded product. The third criterion refers to required throughput. Having established the three criteria, it becomes possible to ascertain the expected dimensions of the performance parameters of the shredder to be designed or selected. Among the operating parameters are rotor rotational velocity, motor power rating, grate spacing or exit clearance, and number of hammers. Performance parameters include energy required per unit weight of material for a given degree of size reduction, hammer wear per weight of material shredded, and magnitude of size reduction.

2. Hammer Tip Velocity

Angular velocity of the shredder rotor—expressed more appropriately as hammer tip velocity—has an important bearing on shredder performance. According to field test data reported by Savage and Shiflett,[22] the hammer tip velocities of the seven horizontal hammermills observed by them were within a range of 55 to 70 m/s. Tip velocities slower than 55 to 70 m/s not only can be, but are applied in the shredding of solid waste. In fact, they describe a Carborundum Model 1000® vertical ring shredder that was operated at a tip velocity of 29 m/s. In two earlier publications, Savage and Trezek[17,24] report investigations in which MSW was successfully shredded at tip velocities of 26 and 38 m/s with the use of a horizontal swing hammermill.

Some evidence exists to the effect that hammer wear is less extensive at relatively low tip velocities (i.e., less than 40 m/s) than at high tip velocities.[24] However, no published data other than those in Reference 24 can be found that either confirm or contradict this finding. An evaluation of the benefits accruing from a reduction in tip velocity must take into account the diminution of the "flywheel effect" of the shredder rotor assembly and flywheel that accompanies the reduction. The diminution is manifested as a lesser storage of energy for shredding transient peak loads. This disadvantage can be minimized by designing additional inertia into the equipment to compensate for the loss due to the reduction in rotational velocity.

The potential of shredding MSW at hammer tip velocities less than 40 m/s is of interest because of the likelihood of an accompanying reduction in hammer wear and of a lessening of the chances of an explosion in the shredder. The possibility of fewer explosions has been postulated by several shredder manufacturers, although as of this writing, no documentation exists to substantiate their claims. The rationale for the "low

Table 6
AVERAGE PRODUCT SIZE AT VARIOUS GRATE OPENINGS AND REFUSE THROUGHPUTS

Minimum grate opening (cm)	Throughput (tons/hr)		Average product size	
	(Metric)	(Short)	Characteristic (cm)	Nominal (cm)
12.7	22.5	24.8	3.7	9.8
22.9	16.9	18.6	5.0	12.7
20.3	35.6	39.3	2.2	6.4
24.1	74.4	82.0	3.2	9.2

Modified from Savage, G. M. and Shiflett, G. R., Processing Equipment for Resource Recovery Systems. III. Field Test Evaluation of Shredders, final report by Cal Recovery Systems, Inc., for U.S. EPA, Cincinnati, Ohio, 1980.

rpm-fewer explosions" hypothesis is the lessening of the generation of sparks within a shredder that accompanies a reduction in tip velocity. Because of the complicated processes involved in shredding solid waste, however, the complete explanation is not likely to be that simple.

3. *Desired Particle Size*

Grate bar spacing can be used only as a very rough guide in the determination of the particle size range of the product. For example, rate of throughput has an effect on the size of the shredded product. Although specific governing relationships had been developed for particular shredders at the time of this writing,[17,25] a universally applicable governing relationship between product size, size of grate opening, and throughput had as yet not become apparent.

Test data reported by Savage and Shiflett[22] and reproduced in Table 6 indicate the absence of a correlation between product size and the dimension of grate openings. Interestingly, the characteristic particle size of the product obtained with the grate opening set at 12.7 cm and the throughput at 22.5 Mg/hr was quite comparable to that obtained with a 24.1-cm opening and a 74.6-Mg/hr throughput, namely 3.2 cm and 3.7 cm, respectively. Apparently, a similar product size was attained by doubling (approximately) the size of grate opening and increasing the throughput by 230%.

4. *Power and Energy Requirements*

Field test data[17,22,25] provide a basis for the development of empirical relationships useful in the ascertainment of the average net specific energy required to shred raw MSW to a given particle size. The data were collected in test runs conducted at seven resource recovery facilities in the U.S. which collectively included nine shredders and shredder capacities ranging from 9.1 to 91 Mg/hr. The relationships developed from the data are Equations 25 and 26, namely,

$$E_o = 17.91\ X_o^{-0.90}$$

or

$$E_o = 35.55\ X_{90}^{-0.81}$$

where E_o is the average net specific energy (kWh/Mg) on an as-processed weight basis; X_o is the characteristic product size in cm; and X_{90} is the nominal product size in cm.

The average gross power requirements for size reducing MSW can be estimated with the use of Equations 1 and 2, of certain basic identities, and of elementary algebra. By definition, the gross power (P_G), net power (P_N), and freewheeling power (P_{FW}) are related by way of the following equation:

$$P_G = P_N + P_{FW} \tag{35}$$

where

$$P_N = E_o\dot{m} \tag{36}$$

E_o is the average net specific energy (kWh/Mg); $\dot{m}$ is the mass throughput (mg/hr) on an as-processed weight basis; and the freewheeling power, P_{FW}, can be expressed as

$$P_{FW} = \alpha P_G \tag{37}$$

where α is the ratio of P_{FW} to P_G, and generally is within the range of 0.5 to 0.20.[22] Substitution of Equations 36 and 37 into Equation 35 yields

$$P_G = E_o\dot{m} + \alpha P_G \tag{38}$$

Through a rearrangement of the terms, the gross power required for size reduction can be expressed in terms of net specific energy and mass throughput as follows:

$$P_G = \frac{E_o\dot{m}}{(1 - \alpha)} \tag{39}$$

Since an empirical relationship exists between the average net specific energy requirement and the nominal product size that is

$$E_o = 35.55 X_{90}^{-0.81} \quad \text{(from Equation 26)} \tag{40}$$

then Equation 39 can be rewritten as

$$P_G = \frac{35.55 X_{90}^{-0.81} \dot{m}}{(1 - \alpha)} \tag{41}$$

or alternatively, P_G can be expressed in terms of the characteristic product size, X_o, using Equation 25,

$$P_G = \frac{17.91 X_o^{-0.90} \dot{m}}{(1 - \alpha)} \tag{42}$$

5. Design Considerations

Equations 25 or 26 can be used to ascertain the average net specific energy required to size reduce raw MSW to a given particle size. The average net power requirement of the shredder's prime mover (e.g., electric motor or diesel engine) is determined by multiplying the average net specific energy by the throughput. To ascertain the average gross power requirement of the prime mover, a freewheeling power contribution must be added to the average net power required as shown in Equations 35 and 38. Savage and Shiflett[22] found the freewheeling power requirements of shredders driven by electric motors typically to be from 5 to 20% of the power rating of the motor. The average gross power required to produce a given characteristic or nominal product size is determined according to Equations 41 and 42, respectively.

Since Equations 41 and 42 represent estimations of the average gross power consumption in size reduction, in actual practice a suitable service factor would be applied to P_G. The service factor is mostly a function of the composition of the waste and of the rotary inertia (Wk^2) of the shredder. In the absence of definitive results to the contrary, it can be reasonably assumed that a service factor of 1.2 to 1.7 would be in a range within which an undue burden would not be placed upon the shredder prime mover during throughput variations characteristic of the loading and conveyance of MSW. However, it should be noted that service factors within this range, i.e., 1.2 to 1.7, are only a little more than 50% of the 2.8 that was average for the shredders observed by Savage and Shiflett in field studies.[22] Whether or not the observed high service factors were a consequence of deliberate overdesign or of an inadequate material throughput is a matter of conjecture at this writing. Finally, it should be emphasized that the 1.2 to 1.7 range applies to the primary shredding of typical MSW, and excludes the shredding of heavier or bulkier wastes, as for example, automobiles, refrigerators, and water heaters.

If the service factor is defined as η, Equations 41 and 42 can be rewritten to express the average design power requirement, P_D,

$$P_D = \eta P_G \tag{43}$$

or

$$P_D = \frac{\eta 17.9\ X_o^{-0.90}\ \dot{m}}{(1 - \alpha)} \tag{44}$$

and

$$P_D = \frac{\eta 35.55\ X_{90}^{-0.81}\ \dot{m}}{(1 - \alpha)} \tag{45}$$

For a range of 1.2 to 1.7, the equations for P_D (44 and 45, respectively) can be recast as

$$21.49\ \frac{X_o^{-0.90}\ \dot{m}}{(1 - \alpha)} \leq P_D \leq 30.45\ \frac{X_o^{-0.90}\ \dot{m}}{(1 - \alpha)} \tag{46}$$

and

$$42.66\ \frac{X_{90}^{-0.81}\ \dot{m}}{(1 - \alpha)} \leq P_D \leq 60.44\ \frac{X_{90}^{-0.81}\ \dot{m}}{(1 - \alpha)} \tag{47}$$

If an average value of α is chosen as 0.10, then Equations 46 and 47, respectively, can be further simplified to

$$23.88\ X_o^{0.90}\ \dot{m} \leq P_D \leq 33.83\ X_o^{-0.90}\ \dot{m} \tag{48}$$

and

$$47.40\ X_{90}^{-0.81}\ \dot{m} \leq P_D \leq 67.15\ X_{90}^{-0.81}\ \dot{m} \tag{49}$$

Equations 48 and 49 express the range of design power for the prime mover of the shredder as a function of the product particle size in the primary size reduction of raw MSW. Either one of the two equations can be used to determine the size of the prime mover. The choice depends upon whether or not the designer prefers to use characteristic instead of nominal product size as the criterion governing design, or vice versa. With respect to estimating power, probably characteristic size would be a good

Table 7
CALCULATION OF SHREDDER POWER REQUIREMENTS

Design Criteria

Material: raw MSW with no greater than typical percentage of oversize bulky wastes
Throughput ($\dot{m}$): 50 metric tons (Mg)/hr
Desired particle size: 90% cumulative weight percent passing 8 cm (X_{90})
Freewheeling power to gross power ratio (α): 0.10
Service factor (η): 1.5

Power Calculations

Gross power, P_G, required: (See Equation 41)

$$P_G = \frac{35.55\ X_{90}^{\ -0.81}\ \dot{m}}{1 - \alpha}$$

$$= \frac{35.55(8)^{-0.81}(50)}{1 - 0.1}$$

$$= 366\ \text{kW}$$

Design power, P_D, required:

$$P_D = \eta\ P_G$$
$$= 1.5\ P_G$$
$$= 550\ \text{kW}$$
$$= 737\ \text{horsepower (hp)}$$

choice, because with it a better correlation with power requirement can be made than can be done with nominal product size.[27] A sample calculation of the shredder power requirements for a specific case is shown in Table 7.

Selection of the precise value for the design power within the estimated range is a matter of experience acquired in the course of shredding a variety of types of MSW. A good rule of thumb is that if actual test data are not at hand, it is prudent to err on the higher side of the power range.

6. Maintenance

Maintenance should be an extremely important consideration in the selection of a shredder. Consequently, an awareness of the many and various maintenance problems that beset a shredder operation is an absolute prerequisite to the intelligent interpretation of the several shredder designs currently available.

The labor-intensive nature of most shredder maintenance constitutes an important element in the costs of shredding. Therefore, the possession of features that permit or promote the performance or maintenance in an expeditious manner should be an important criterion in the selection of a shredder. Among such features are the following five:

1. Appropriate equipment (e.g., overhead cranes, hydraulic or pneumatic devices for opening the shredder for access to the internal parts and for closing it, mechanical pin pullers, hammer removal jigs, and reversible hammers)
2. Appropriate and well-kept welding equipment and facilities that include a well-designed ventilation system

3. The lighting required for the safe performance of maintenance operations
4. Movable scaffolding or permanent walkways about the shredder to permit easy access to all pieces of equipment that may require maintenance
5. Adequate space around the shredder to accommodate the removal and reinstallation of grate bars, hammers, wear plates, breaker bars, and other parts

Item 4 includes sufficient crawlways and walkways below and adjacent to the shredder and its auxiliary equipment, especially the conveyors.

To be complete, the specifications for a shredding system should call for all five features. Moreover, the solicitation for price quotations should be accompanied by the stipulation that the responding manufacturer specify all equipment and material inventory needed in the maintenance of his shredder. The latter equipment includes that required for bearing and rotor replacement or overhaul, and for the removal and reinstallation of hammers, liners, wear plates, breaker bars, and of grate bars or cages.

D. Operational Alternatives

1. Multiple-Stage Size Reduction

Coarse or primary shredding, grinding, or milling, usually to a nominal particle size of about 10 cm, is a feature of practically all resource recovery facilities. Secondary and even tertiary shredded (collectively termed "fine" shredding herein) are introduced whenever a particle size significantly smaller than 10 cm is specified, as for example in the production of refuse-derived fuel (RDF). Thus, the particle sizes of RDF "fluff" usually are within the range of 90% passing 2.5 to 7.5 cm. For efficient and reliable densification, the particle size of feedstock for RDF pelletization usually must be in the range of 1 to 2.5 cm. Other applications in which fine shredding may be involved are the recovery and processing of ferrous, aluminum, and glass scrap to meet user specifications.

2. "Separate Grinding"

The term "separate grinding" as used here refers to size reduction of the low density, pliable components (e.g., paper, plastics) of MSW separately from that of its brittle components (e.g., glass, metal).

Experience in general indicates that the efficiency of the size reduction process can be significantly increased by separate grinding. Although several factors are responsible for the increase, suffice it to point out here that they center around the fact that in the size reduction of low density, pliable components, the specific energy consumption and the wear-and-tear on the shredder generally are greater than in the size reduction of the high density brittle components.

Separation of refuse into the two constituent groups (low density pliable and high density brittle) assumes an especial importance in fine size reduction. Because fine grinding is an expensive operation in terms of energy consumption, any means of reducing that consumption warrants attention. Since separate grinding is one such means, separation of refuse into its two constituent types certainly would be a useful prelude to fine grinding.

3. Fine Grinding

Results obtained in research by Savage and Shiflett[22] give reason for concluding that the use of the multistage approach to fine size reduction may be more efficient than single-stage grinding with respect to energy consumption. On the other hand, data reported by Shiflett and Trezek from other research[20] lead to quite the opposite conclusion; namely, that the more efficient approach would be a single-stage size reduction in which a virtually choked primary shredder is used. The existence of data that justify two apparently diametrically opposed, albeit tentative conclusions, most likely is explained by the many unknowns associated with key machine parameters, especially those related to grate spacing, number and shape of hammers, and the internal volume of the shredders. Until these machine parameters and their influence on the mechanisms of size reduction are studied and more fully understood, no definitive answer can be given as to the validity of either of the two conclusions. Certainly, each has some justification. In the absence of a final answer, the best course to follow probably would be one involving a coarse size reduction of the incoming raw MSW in a single-stage operation. This would be followed by processing sequences designed to separate the high density brittle materials from the low density pliable materials, and finally, by a secondary shredding of either or both of the segregated streams.

Separation of the refuse into its low density and high density components can be made prior to coarse shredding through the use of a trommel or other device. It can also be done after shredding either through air classification or through a combination of air classification and screening. Regardless of which of the two approaches is followed, the outcome is a separation of the incoming waste stream into one fraction that contains most of the high density materials (i.e., trommel undersize or air classifier heavies), and into another that contains the low density, mostly pliable materials (i.e., trommel oversize or air classifier lights). Consequently, if further size reduction is required, the materials in each of the two streams can be size reduced in a shredder more or less specifically designed to process them.

REFERENCES

1. **Austin, L. G. and Klimpel, R. R.,** Theory of grinding operations, *I and EC Process, Design and Development,* 56, 19, 1964.
2. **Snow, R. H.,** Annual review of size reduction, *Powder Techn.,* 5, 351, 1971/72.
3. **Fuerstenau, D. W. and Somasundaran, P.,** Cinetique de ta fragmentation, in *6th Congres International de Ta Preparation des Minerals (Cannes),* Saint-Etienne, Ed., 1963, (see also "Comminution kinetics," in *Mineral Processing,* Roberts, A., Ed., Pergamon Press, Oxford, 1965, 25.
4. **Kinasevich, S. R.,** Application of the Schumann single event hypothesis to comminution, M.S.thesis, University of California, 1962.
5. **Charles, R. J.,** Energy-size reduction relationships in comminution, *Trans. AIME,* 208, 80, 1957.
6. **Holmes, J. A.,** Contribution to the study of comminution—modified Kick's law, *Trans. Inst. Chem. Engrs.,* 35, 125, 1957.
7. **Schumann, R.,** Energy input and size distribution in comminution, *Trans. AIME,* 217, 22, 1960.
8. **Gaudin, A. M.,** An investigation of crushing phenomena, *Trans. AIME,* 73, 253, 1926.
9. **Schumann, R.,** Principles of comminution I. Size distribution and surface calculations, *Tech. Publs. AIME,* 1189, 11, July 1940.
10. **Rosin, P. and Rammler, E.,** Laws governing the fineness of powdered coal, *Inst. of Fuel,* 729, 1933.
11. **Gaudin, A. M. and Meloy, T. P.,** Model and a comminution distribution equation for single fracture, *Trans. AIME,* 223, 40, 1962.
12. **Bergstrom, B. H.,** Empirical modification of the Gaudin-Meloy equation, *Trans AIME,* 235, 45, 1966.
13. **Callcott, T. G.,** A study of the size reduction mechanisms of Swing Hammer Mills, *J. Inst. Fuel,* 33, 529, 1960.
14. **Obeng, D. M.,** Comminution of a Heterogeneous Mixture of Brittle and Non-Brittle Materials, Ph.D. dissertation, University of California, Berkeley, 1973.

15. **Broadbent, S. R. and Callcott, T. G.,** A matrix analysis of processes involving particle assemblies, *Phil. Trans. Roy. Suc.,* 249, Series A, 99, 1956.
16. **Savage, G. M., Trezek, G. J., and Shiflett, G. R.,** *Size reduction in solid waste processing—fine grinding.* Progress Report 1976-1978, Report to the U.S. EPA, prepared under Grant No. EPA R803034.
17. **Trezek, G. J. and Savage, G. M.,** *Significance of size reduction in solid waste management,* EPA-600/2-77-131, Municipal Research Laboratory, Office of Research and Development, U.S. EPA, Cincinnati, Ohio, 1977, 149.
18. **Trezek, G. J. and Savage, G. M.,** *Size reduction in solid waste processing,* Progress Report 1973-1976, Report to U.S. EPA, prepared under Grant No. EPA R-80128, 1976.
19. **Shiflett, G. R.,** A Model for the Swing Hammermill Size Reduction of Residential Refuse, D. Eng. dissertation, University of California, Berkeley, 1978.
20. **Shiflett, G. R. and Trezek, G. J.,** Parameters governing refuse comminution, *Res. Rec. Conserv.,* 4, 31, 1979.
21. **Gawalpanchi, R. R., Berthovex, P. M., and Ham, R. K.,** Particle size distribution of milled refuse, *Waste Age,* 4(5), 34, 1973.
22. **Savage, G. M. and Shiflett, G. R.,** *Processing equipment for resource recovery systems, III. Field test evaluation of shredders,* Final Report by Cal Recovery Systems, Inc., for U.S. EPA, Cincinnati, Ohio 1980.
23. Summary of Project and Findings, Preparation of RDF on a Pilot Scale, National Center for Resource Recovery, June 1977.
24. **Savage, G. M. and G. J. Trezek,** On grinder wear in refuse comminution, *Compost Sci.,* 15 4, 1974.
25. **Savage, G. M., G. R. Shiflett, and G. J. Trezek,** *Significance of Size Reduction in Solid Waste Management—Part II,* prepared for the Environmental Protection Agency, 1979.

Chapter 5

AIR CLASSIFICATION

I. INTRODUCTION

A. Principles

For the present, air classification finds its use in resource recovery principally as a means of separating from the solid waste stream the fraction that is rich in combustible materials. Essentially, air classification depends upon the interaction between a moving air stream and shredded refuse within a column. This interaction leads to the generation of drag forces on the particles that are simultaneously opposed by the gravitational force, which also is acting upon the particles. As a result of the interplay of the two forces exerted on the refuse particles within the column, refuse components characterized by a large drag-to-weight ratio are suspended in the air stream. On the other hand, components that have a small drag-to-weight ratio tend to settle out of the stream. The suspended fraction is conventionally termed "air-classified light fraction", or simply as "lights". Not unexpectedly, the settled fraction is known as the "air-classified heavies", or simply as "heavies". The unit in which the air classification takes place is designated an "air classifier".

In air classification, paper and film plastic tend to be concentrated in the light fraction. Because of this concentration, most of the materials in the light fraction have a significantly high heating value, and the terms "light fraction" and "combustible fraction" often are used interchangeably, albeit incorrectly because strictly speaking, the terms are not synonymous. They are not synonymous because the lights include a significantly large percentage of materials that are not combustible. Metals and glass constitute the principal components of the heavy fraction. Because of the physical principles that govern the process of air classification, materials other than metals and glass may end up in the heavies fraction. The reasons for the less than precise separation indicated in the preceding statements will become apparent in the discussion on the aerodynamics of air classification in a later section of this chapter.

The extent of the separation to be expected of air classification is illustrated by the experimental data listed in Tables 1 and 2. With a municipal solid waste having the composition indicated by the data in Table 1 and size reduced to a nominal particle size of −5 cm, the compositions of the lights and heavies would typically be comparable to those listed in Table 2. A perusal of the data in the two tables will show that while the combined concentrations of metals and glass in the raw refuse in this instance amounted to 18.8%, those of the heavy fraction after separation in the air classification process was 62.6%. The differences between concentrations of combined paper and plastics were somewhat less drastic, being on the order of 47.7% and 65.9%, respectively in the raw refuse and the light fraction.

A number of air classifiers are in various stages of operation in the U.S. A partial list of the operations is given in Table 3.

II. TYPES OF AIR CLASSIFIERS

Despite the considerable number of air classifier designs in existence, they are all based upon identical fundamental principles and can be fitted into one of three broad groups, namely horizontal, vertical, and inclined.

While as of this writing at least one manufacturer offers a horizontal type and two

Table 1
COMPOSITION OF A TYPICAL REFUSE[a]

Component	Weight (%)
Fe	7.3
Glass	10.8
Aluminum	0.7
Paper	43.2
Plastic	4.5
Miscellaneous	33.5
	100.0

[a]Unpublished University of California data.

Table 2
COMPOSITIONS OF THE AIR-CLASSIFIED HEAVY AND LIGHT FRACTIONS

Heavies		Lights	
Component	Weight (%)	Component	Weight (%)
Fe	24.2	Paper	60.2
Glass	36.1	Plastic	5.7
Aluminum	2.3	Miscellaneous	34.1
Paper	3.6		
Plastic	1.7		
Miscellaneous	32.1		
	100.0		100.0

Modified from Trezek, G. and Savage, G., msw component size distributions obtained from Cal resource recovery system, *Resour. Recovery Conserv.*, 2, 67, 1976.

manufacturers offer an inclined type of air classifier, the greater majority of the existing classifiers are vertically oriented. In the horizontal type, the air stream moves through a horizontal section before being drawn out either vertically or horizontally from the classifier. The design of a horizontal classifier is shown in Figure 1. The design of an inclined air classifier contains features of both the vertical and horizontal type of air classifier as shown in Figure 2. Inclined air classifiers are taken to include both the vibrating and rotary types. In a vertical classifier, the air stream moves vertically through the classifier column in the manner shown in Figure 3. The designs of vertical classifiers may vary among themselves in several ways. For example, baffles ("zig-zags") may be inserted in the column, or the cross-sectional geometry may be either circular or rectangular. Other variations may be in the method of feeding shredded refuse into the air column (e.g., by way of chute or by rotary airlock), or of removing the light and heavy fractions from the column.

Regardless of design, the upper limit of throughput capacity presently is around 70 Mg (80 tons)/hr. Units with higher design capacities have been built, but they have never been operated successfully. Nevertheless, as the push towards 907 Mg/day (1000 tons/day) resource recovery facilities begins to take place, the working capacity may

Table 3
LOCATION AND DESCRIPTIONS OF AIR CLASSIFIERS OPERATED IN THE U.S.

Facility	Location	Nominal through-put (Mg/hr)	Unit description	1979 Mode/status
Ames Resource Recovery System	Ames, Iowa	27	Vertical straight column	Production/ operational
St. Louis Refuse Processing Plant	St. Louis, Mo.	27	Vertical straight column	Production/ nonoperational
Recovery I	New Orleans, La.	36	Vibrating inclined column	Production/ nonoperational
NCRR	Washington, D.C.	5	Vibrating inclined column	Research/ nonoperational
Lane County	Portland, Ore.	54	Vertical straight column	Production/ shakedown
Vista Chemical	Los Gatos, Calif.	18	Vertical zig-zag column	Research/ operational
City of Akron	Akron, Ohio	54	Vibrating inclined column	Production/ shakedown
Pompano Beach	Pompano Beach, Fla.	5	Vibrating inclined column	Production/ operational
University of California	Richmond, Calif.	4	Vertical straight column	Research/ operational
Baltimore County	Cockeysville, Md.	18	Vertical tube	Production/ operational
Monroe County	Rochester, N.Y.	54	Inclined rotating drum	Production/ shakedown
Tacoma	Tacoma, Wash.	73	Horizontal column	Production/ shakedown
City of Los Angeles	Los Angeles, Calif.	4	Vertical straight column	Research/ operational
Chicago Southwest	Chicago, Ill.	36	Vibrating inclined column	Production/ non-operational

indeed be brought up to 115 Mg (125 tons)/hr. Before that happens, however, a number of significant and substantial problems in design and scale-up must be solved.

Because the vertical type of classifier is the one most commonly encountered, it is the type that receives the major attention herein. The advisability of focusing attention on the vertical form is confirmed by the fact that the greater part of the experimental data thus far reported in the literature deals almost exclusively with vertical air classification. In fact, neither operational track records nor reports based on objective testing and analytical evaluations of air classifiers other than of the vertical type were available.

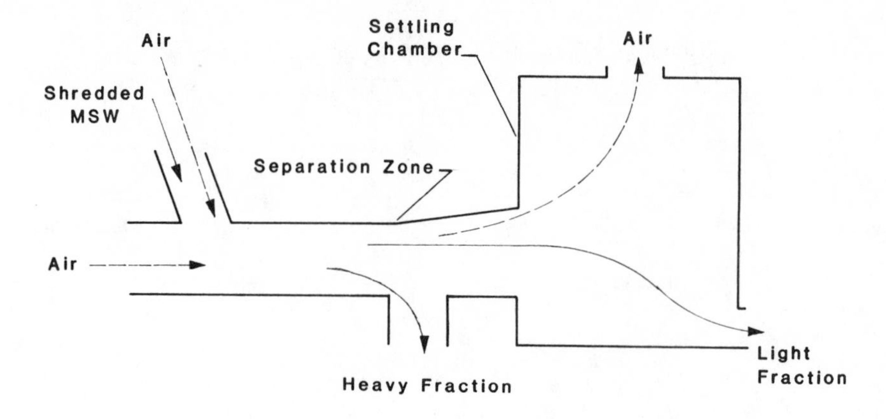

FIGURE 1. Horizontal air classifier.

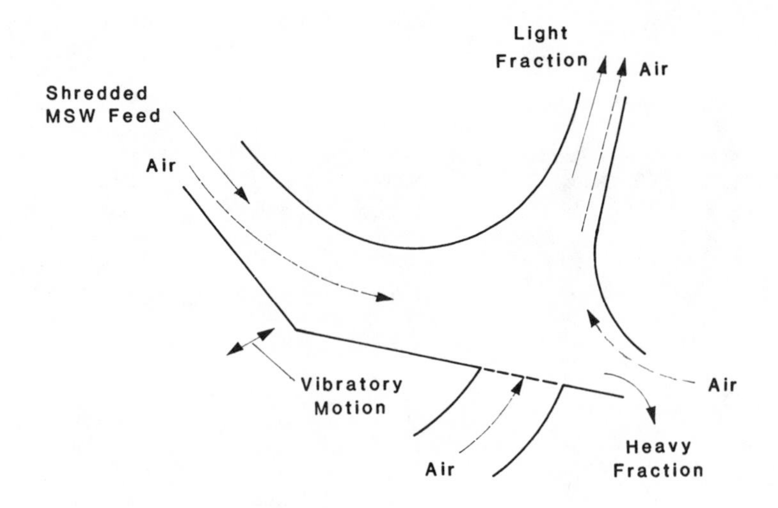

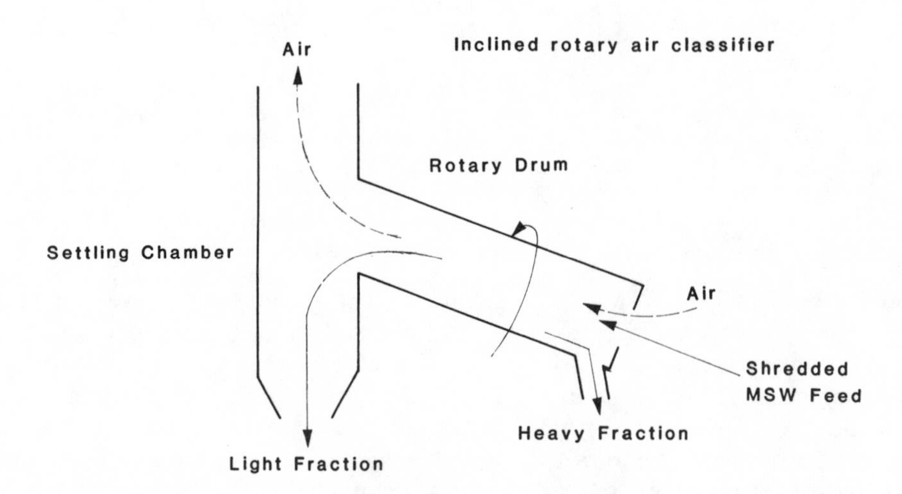

FIGURE 2. Inclined air classifier.

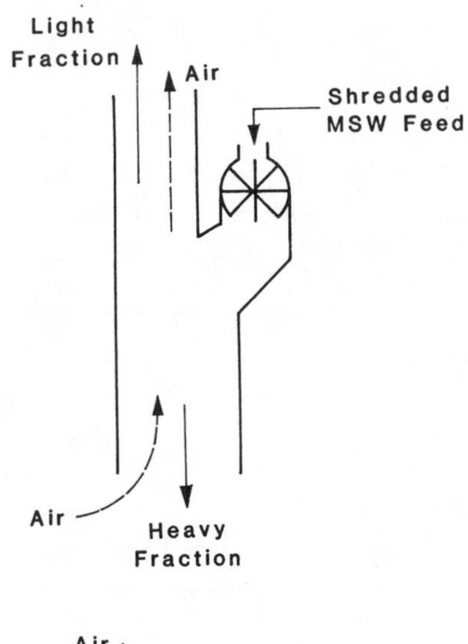

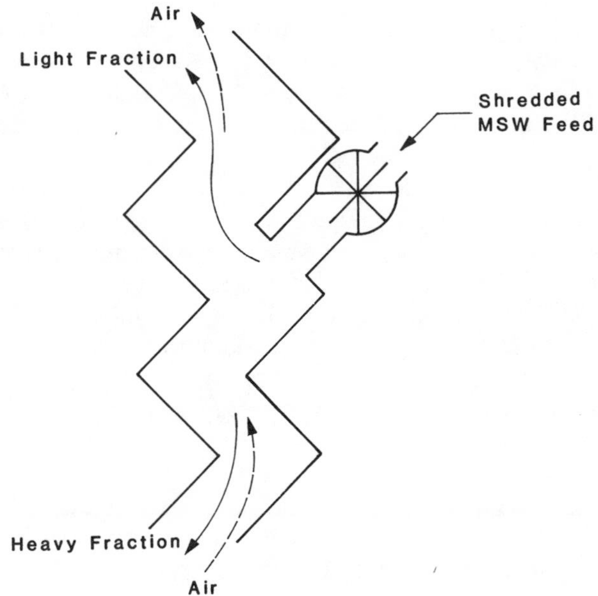

FIGURE 3. Vertical air classifier.

A point in favor of the vertical air classifier is the ease with which it can be modeled, inasmuch as the unit generally would be in the form of a straight column. On the other hand, modeling an inclined or rotary air classifier would be a rather complex problem. However, its greater complexity does not necessarily mean that the performance of an inclined or rotary air classifier would be less satisfactory than that of the vertical type.

Finally, it should be noted that the applicability of the information disclosed in this chapter can be extended to all types of air classifiers under the premise that all air classifiers conform to identical governing principles.

III. DESIGN AND OPERATIONAL FACTORS

A. In-Feed Systems

Although ideally an in-feed system to an air classifier should provide a uniform rate of feed, many factors interfere with the achievement of that uniformity. First and foremost is the variability of the composition and density of refuse, even from point to point within a given system. Despite the fact that the vibrating conveyor is a very versatile piece of equipment for providing relatively uniform flows, its use in handling shredded MSW has been met with very limited success. Furthermore, it is a rather expensive means of providing in-feed control.

The rotary valve airlock is another candidate system. However, it has an inherent disadvantage in that its basic design results in an intermittent feeding of material into the air classifier. Moreover, the unit is susceptible to jamming if the clearance between the rotating vanes and the outer housing is not sufficiently large, and if the motor power is not great enough to dislodge material that might jam or compact between the rotary vanes and the housing. If a rotary valve airlock is used, provision must be made for the storage of material upstream of the feeder. The reason for the provision ultimately is a lack of capacity of the part of the unit to accommodate surges of throughput material greater than its maximum capacity. Airlock feeders operate at a given rotational velocity, and, therefore, fixed volumetric flow rate. Consequently, they cannot feed material at a rate in excess of that permitted by the device's maximum volumetric displacement per unit time. Consequently, a surge in excess of that maximum must be accommodated upstream of the feeder. If such an accommodation is not available, then the maximum capacity of the feeder must *never* be exceeded.

Strictly speaking, the third method to be mentioned at this time is not a method, but rather is a twofold design provision for a pseudo-airlock "seal" and a column cross-section sufficiently large to accommodate surges of infeed material. Usually, the "seal" takes the form of a curtain or series of curtains that span the air classifier in-feed opening, and drag over the in-feed material being transported on the conveyor belt. The curtain, typically of rubber, functions as a pseudo-airseal.

B. Air Classifier Split

"Air classifier split" is a term commonly used to describe the division of the input material into the light and the heavy fractions on a mass-flow basis. For a given mass-flow rate of refuse ($\dot{m}_s$), continuity requires that

$$\dot{m}_s = \dot{m}_{HF} + \dot{m}_{LF} \quad (1)$$

where $\dot{m}_{HF}$ and $\dot{m}_{LF}$ are the mass-flow rates of heavies and of the lights, respectively. The split generally is reported on a mass fraction basis, with the heavies split being equal to $\dot{m}_{HF}/\dot{m}_s$, and the light split to $\dot{m}_{LF}/\dot{m}_s$. Splits can be based upon dry weight, air-dry weight, or wet weight. Wet weight is the basis most commonly relied upon in the industry.

Air classifier split can also be expressed as the ratio of the light fraction to the heavy fraction (e.g., 80/20, 70/30). The sum of the mass fractions of heavies and lights obviously must be equal to unity.

The air classifier split depends upon several variables, among which are air-to-solids ratio, column velocity, refuse moisture content, and column loading. To determine the nature of a classifier split, a simultaneous measurement of heavy and light fractions must be made over a given time period under constant, or steady-state, operating conditions. Preferably the entire outputs of the heavy and light streams should be collected within the sampling time interval. Results can be reported on any of the mass bases described in the two preceding paragraphs.

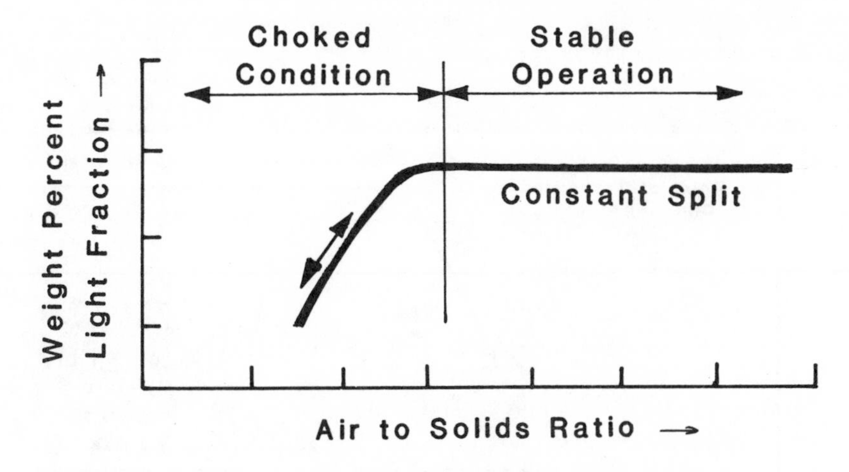

FIGURE 4. Relation between efficiency and air-to-solids ratio.

C. Material Loading (Air-to-Solids Ratio)

The material loading, or air-to-solids ratio, is a very important design parameter in air classification inasmuch as a minimum ratio exists below which effective separation of lights and heavies ceases to occur within the air classifier column. At a given air velocity and columnar cross-section (i.e., air flow rate), a drop in efficiency of separation is manifested by an increase in the amount of combustible (low density) material appearing in the heavy fraction, and a corresponding decrease in that carried away with the light fraction. The relation between efficiency and air-to-solids ratio is illustrated by the curve in Figure 4. The abrupt transition from variation to stability indicated in the figure apparently was found to be characteristic of all of the vertical classifiers observed by the authors of this book.

Among the several factors that determine the minimum material loading value are (1) internal geometry of the air classifier; (2) moisture content of the shredded refuse; (3) composition of the refuse; (4) particle size spectrum. With a municipal refuse shredded to −10 cm, the first and second factors constitute the dominant parameters. For example, in a study involving a vertical air classifier that had a rectangular cross-sectional area of 0.4 m^2, and in which the air flow rate was 2.6 m^3/s and the column velocity was 7.0 m/s,[1] the split dropped from 84/16 to 73/27 when the bulk density of the input material was increased from 104 to 144 kg/m^3. (In the particular instance, the increase in bulk density was a function of increase in moisture content of the refuse.) The effect is illustrated graphically by the differences in the steady-state values of the curves A and C in Figure 5. Inasmuch as the column air velocity remained constant, the greater density of the input material brought about an increase in the weight-to-drag ratio. The upshot was that more combustible material dropped out with the heavy fraction.

There is evidence that the point of transition between variation and stability in light fraction separation (indicated by the "knee" of the curve in Figures 4 and 5) may be a function of both the bulk density of the input material and the air velocity within the column. As far as density is concerned, its effect is demonstrated by the shift to the left of the transition point in curves A and B in Figure 5. The two curves represent the denser material. The effect of column velocity is shown by the difference between curves A and C in the same figure. In the latter case, the transition point has shifted to the left at the lower velocity. In fact, the transition point for curve C is beyond the leftmost limit of the experimental data. The two curves also illustrate the effect of

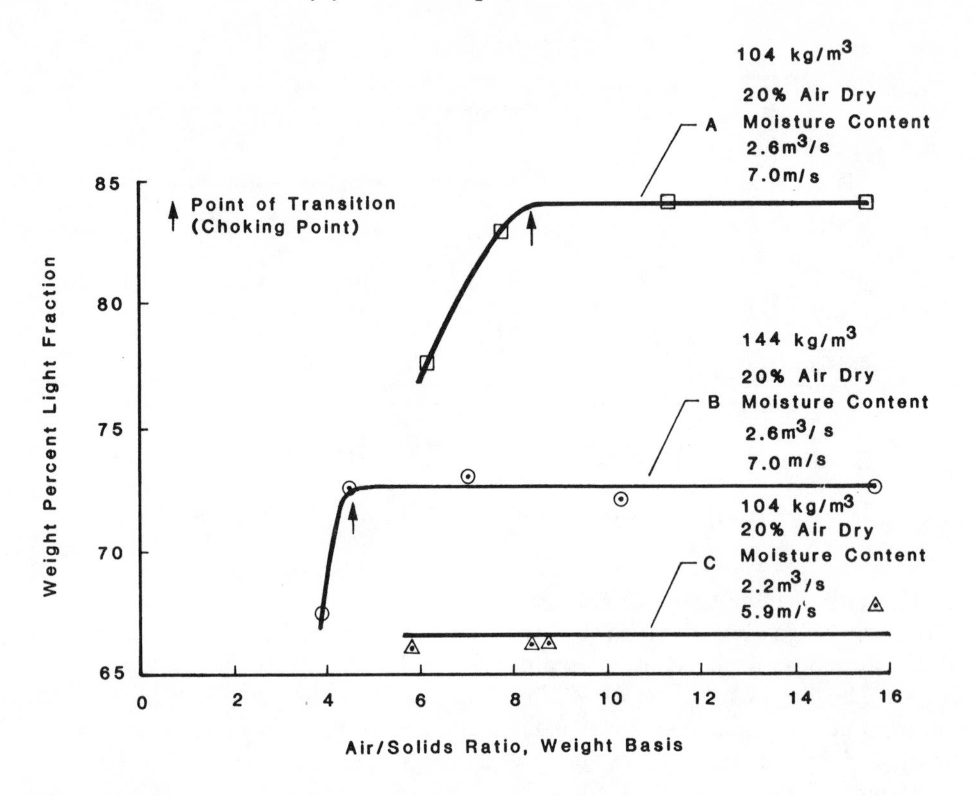

FIGURE 5. Effect of operating conditions on the light fraction split by a vertical air classifier.

column velocity upon constant split, i.e., indicated by the horizontal portion of the curve on the right of the transition point. As curves A and B show, the percentage of wastes in the separated light fraction dropped from 84% to 67% when the velocity was reduced from 7.0 to 5.9 m/s.

The basic relationship of the position of the transition point or "knee" of the material-loading curve as a function of the air-to-solids ratio is masked by the interfering effects of bulk density and column velocity. Therefore, a concerted research effort must be made in the area of material loading and its governing parameters before a rational design philosophy for air classifiers can be developed. Until that is done, a review of the material loading values used in the refuse processing industry will have to serve as a substitute for basic design information.

In the several air classification operations in which the authors have participated or had contact, the air-to-solids ratios applied ranged from 1.4 to 7.4, inclusively. Data pertaining to these operations are listed in Table 4. Best performance was obtained with the University of California pilot-scale air classifier when the ratios were within the range of 3.9 to 7.4:1. Murray[2,3] and Murray and Liddell,[4] working with a pilot-scale Allis Chalmers Dual Vortex Air Classifier®, noted that the ratio could be dropped to a low of 2.4:1 before the transition point ("knee") in the material loading curve was reached. Despite their experience, more experimental evidence is needed before it would be prudent to operate the air classifier at an air-to-solids ratio less than 2:1.

Inasmuch as the attainment of maximum separation efficiency depends upon the maintenance of an air-to-solids ratio above a certain minimum level, and that minimum is a function of the given set of operating conditions, it follows that a method for estimating the minimum ratio can be used to good advantage. One such method is as follows:

Table 4
SUMMARY OF OPERATING PARAMETERS FOR VERTICAL AIR CLASSIFIERS USED AT VARIOUS SITES

Site	Split (LF/HF) (%)	Column velocity (m/s)	Solids flow rate (Mg/hr)	Air-to-solids ratio	Column loading ([Mg/hr]/m)
Ames[a]	84.1/15.9	16.6	29.5	2.6	28.1
St. Louis[b]	81.7/18.3	10.6	31.4	1.9	24.8
Univ. of Calif.[c]	70.0/30.0	7.6	1.4	3.9	6.7
Univ. of Calif.[c]	70.0/30.0	5.3	1.4	5.0	6.7
Univ. of Calif.[c]	84.0/16.0	7.0	1.5	7.4	4.0
Univ. of Calif.[c]	73.0/27.0	7.0	2.5	4.6	6.7
Appleton[d]	79.6/20.4	9.1	4.5	2.4	17.9
Appleton[d]	66.9/33.1	9.1	4.5	2.4	17.9

[a] Abstracted from Even et al.[5]
[b] Abstracted from Fiscus.
[c] Unpublished U.C. data.
[d] Data from Murray and Liddell.[2,4]

If by definition

$$\alpha_{min} \equiv \frac{\dot{m}_o}{\dot{m}_s/_{min}} \quad (2)$$

then

$$\dot{m}_o = \alpha_{min}\dot{m}_s \quad (3)$$

the volumetric flow of air is

$$q_o = \frac{\dot{m}_o}{\rho_o} \quad (4)$$

Substitution of Equation 2 into Equation 3 yields

$$q_{o_{min}} = \frac{\alpha_{min}\dot{m}_s}{\rho_o} \quad (5)$$

where α_{min} is the minimum air-to-solids ratio as determined through experiment; $\dot{m}_o$, mass flow of air per unit time; $\dot{m}_s$, mass flow of solids per unit time; ρ_o, density of air; and q_o, volumetric flow of air per unit time.

The minimum air-to-solids ratio for constant split, as described by the "knee" of the material loading curve (Figure 4), marks the point at which the upward flow of air is disturbed to the extent that the lights no longer are buoyed, and therefore begin to settle. For want of a better one, the term "choking point" is given to this point. Thus, the minimum air-to-solids ratio for a constant split in an air classifier is the "choking point" for that unit.

D. "Choked" Condition

As indicated in the previous section, the onset of the "choked" condition is signified by a marked increase in the percentage of material discharged as the heavy fraction. Reports dealing with research on the circumstances that lead to the choked condition are scant. Results of unpublished studies conducted at the University of California

(Berkeley) with the use of a 1-ton and a 4-ton/hr vertical air classifier indicate that choking is primarily a function of the geometry of the classifier column. Apparently, restriction of the air flow due to an overloading of the column with refuse is not a significant factor in "choking". The minor influence of overloading in this respect is evidenced by the fact that even in the choked condition the total air flow is not decreased. In the university studies, a decrease would have been reflected by a constant pressure drop in the system, and such a drop was not detected. An explanation of the relation between "choking" and column geometry awaits further investigation.

The importance of predicting the minimum air-to-solids ratio, i.e., the "choking" point, stems from the desirability of operating an air classifier within its stable range, i.e., when the split of the input refuse has the constancy shown to the right of the knee in Figure 4. To do this, it is necessary that the air-to-solids ratio be beyond the minimum for the classifier. If a classifier is operated while in a choked condition, the split will vary. The nature of the split will depend upon the mass flow rate of refuse into the classifier. Since existing refuse processing systems commonly are characterized by abrupt changes in material flow rate, the continued operation of a classifier in the choked condition results in an unstable process. Such a situation is not conducive to the maintenance of system control.

To keep the operation of an air classifier in the stable mode, a rather large columnar cross-section may be necessary so that a sufficiently high air-to-solids ratio be assured. Inasmuch as the provision of such a large cross-sectional area would be expensive, a trade-off must be sought between stability of performance and economy. Nevertheless, if a control of the air classification process is to be attained, then economy must be sacrificed at least to the degree needed to provide the cross-section required to ensure a stable performance. It may be that presently many air classifiers are being operated in the choked condition due in part to the fact that economics dictates the utilization of smaller than optimum columnar cross-sections.

Beyond the "choking" point, the split of shredded refuse into heavy and light fractions becomes constant and remains independent of the air-to-solids ratio over a wide range, the dimensions of which are determined by the properties of the refuse and operational parameters of the air classifier. Pertinent refuse properties are composition, particle size, and moisture content. Operational factors are volumetric flow rate of air and internal geometry of the column.

When a classifier is operated in the stable mode, perturbations around the operating setpoint, i.e., variations in material flow rate, do not result in sudden overloadings sufficiently severe as to bring about a choked condition, and thereby lead to the discharge of lights with the heavies. However, since the feedrate to an air classifier can vary instantaneously by as much as 100% of a nominal value, prudent design practice calls for the normal operating point to be established well beyond the minimum air-to-solids ratio ("choking" point). The advisability of such a course is demonstrated by the nature of curve B in Figure 5. As the figure shows, in the particular experiment the percentage recovery of lights dropped from 73% down to 68% when the air-to-solids ratio was reduced from 5:1 to 3:1. In the experiment, the reduction in the air-to-solids ratio was roughly equivalent to a 70% increase in solids flow rate.

E. Column Loading

Column loading, that is, material flow rate per unit of air classifier cross-section, is a useful design parameter. Experimental evidence supports the contention that a minimum level of column loading exists below which the quality of classifier performance deteriorates. This minimum level is closely associated with that of the minimum air-to-solids ratio that marks the onset of the "choking" condition. Since among the many classifiers a variety of cross-sectional areas is included, each will have its particular characteristic minimum loading level. Therefore, the minimum loading level becomes a significant design parameter in the determination of the required cross-sectional area for a given air classifier.

According to the data in Table 4, loadings within the range of 4.0 to 28.1 $Mg/m^2/hr$ have been tried. In their studies, the authors of this book observed a satisfactory classifier performance at column loadings of 4 to 7 $Mg/m^2/hr$. The precise loading level was found to depend upon the composition and moisture content of the refuse.

A relation can be developed for the minimum value of column loading (β) at given material flow per unit time ($\dot{m}_s$), column velocity (V_o), and minimum air-to-solids ratio (α_{min}). Since by definition

$$\alpha_{min} \equiv \left.\frac{\dot{m}_o}{\dot{m}_s}\right|_{min} \qquad (6)$$

the volumetric flow of air is

$$\dot{q}_o = \frac{\dot{m}_o}{\rho_o} \qquad (7)$$

Combining Equations 6 and 7 yields

$$\dot{q}_o = \frac{\alpha_{min}\dot{m}_s}{\rho_o} \qquad (8)$$

as well as

$$\dot{q}_o = A_x V_o \qquad (9)$$

Using Equations 8 and 9 to solve for A_x, then

$$A_x = \frac{\alpha_{min}\dot{m}_s}{\rho_o V_o} \qquad (10)$$

Since by definition

$$\beta \equiv \frac{\dot{m}_s}{A_x} \qquad (11)$$

a substitution for A_x using Equation 9 yields

$$\beta_{min} = \frac{\rho_o V_o}{\alpha_{min}} \qquad (12)$$

where α_{min} is the minimum air to solids ratio; $\dot{m}_o$, mass flow of air per unit time; $\dot{m}_s$, mass flow of solids per unit time; ρ_o, density of air; $\dot{q}_o$, volumetric flow of air per unit time; A_x, cross-sectional area of classifier column; V_o, average air velocity; and β_{min}, minimum value of column loading.

The column air velocity, V_o, having been chosen and the minimum air-to-solids ratio determined experimentally, the minimum column loading (β_{min}) can be calculated according to equation 12. The resulting value can then be used to determine the column dimensions required for accommodating the expected material flow rate ($\dot{m}_s$). Thus, scale-up becomes possible due to the uniqueness of the parameter, β_{min}. Of course, this presupposes that the geometrical configuration of the columns remains the same.

F. Pressure Drop Considerations

As with any pneumatic process, pressure losses occur in the process equipment used in air classification. Usually an air classification system consists of an air classifier, ducting, a cyclone for removal of the light fraction, a dust collector, and a blower. Pressure losses are sensitive to velocity of the air stream as well as to the geometry of equipment through which the air flows. Pressure drops by way of ducting, transition pieces, and entrance and exit conditions can be estimated by referring to any number of references on heating, ventilating, and air conditioning (e.g., see Jorgensen[7]). Values for pressure drops across cyclones and dust collectors as functions of air flow can be ascertained by consulting the equipment manufacturers. Reports for values for pressure drops across air classifiers were not to be found in the extensive literature available to the authors of this book. However, a judicious guess would place the range at 0.2 to 2.0 cm of water for a straight column, depending upon internal geometry. This range does not include entrance or exit transition losses, which realistically also would be considered as part of the overall pressure drop in an air classifier.

As a point of reference, air classification systems with capacities of 20 to 40 Mg (22 to 44 tons)/hr can be expected to have total pressure drops within the range of 25 to 50 cm of water. It should be remembered that this range is highly dependent upon the equipment used in the air classification system.

G. Power Consumption

After shredding, the air classification system generally is the most energy intensive process associated with front-end processing. Its significant power requirement results from the relatively large air flows and pressure drops that are a part of air classification. In the resource recovery operation at Ames, an average of approximately 4.1 kWh/Mg is used in an air classification system that processes roughly 30 Mg/hr. In the University of California air classification system (capacity 3.6 Mg/hr) from 3.4 to 4.2 kWh, inclusively, are used per metric ton of throughput. Moreover, the two systems do not include a dust collection system downstream of the cyclone. Obviously, if a baghouse were included, the specific energy (kWh/Mg) would be correspondingly increased because of the additional pressure drop.

H. Yields and Overall Efficiency

A standard is needed in air classification technology whereby the overall efficiency of a given system can be evaluated and its performance be related to its parameters, i.e., α_{min}, β_{min}, V_o, $\dot{q}_o$, $\dot{m}_s$, etc. As of this writing, there is no commonly accepted method of determining classifier performance. A method that has been used to some extent involves a comparison of the amount of aluminum contamination in the light fraction with that in the feed material.[8] In most of the other methods, reliance is had on visual observation for the extent of contamination of light and heavy fractions. Obviously, the visual observation approach is highly subjective and lacks scientific certainty.

A part of the problem in defining an efficiency for air classification stems from the nebulous nature of the definitions devised for the light and the heavy fractions. If these fractions cannot be adequately defined as to composition, then it is equally difficult to numerically characterize their contamination. Inasmuch as paper fiber and plastic normally constitute from 60 to 80% of the light fraction by weight and are the chief materials that contribute to the heating value of the light fraction, it might be said that a high degree of removal of these two materials is tantamount to a high separation efficiency. However, in the process of recovering 100% of the fiber and plastic in the light fraction, certain contaminants principally in the form of noncombustible particles of ferrous, aluminum, dirt, and glass also may be carried over. Thus, even at a full separation of all combustibles, an air classifier that produces a light fraction significantly contaminated with such noncombustibles could hardly be designated as being efficient. With this latter fact in mind, it would be reasonable to describe air classifier efficiency in terms of two quantities: (1) percent recovered paper and plastic (R_{pp}), defined as the weight ratio of paper and plastic in the light fraction as compared to the total paper and plastic in the waste; and (2) retained ash (R_a), defined as the ratio of ash in the light fraction to the total ash of unclassified waste. Using the two terms, an efficient air classifier would be one that recovers a high percentage of paper and plastic and rejects a high percentage of the materials that contribute to ash. For example, the reported yield of paper and plastic obtained with a prototype vertical classifier[2] was such that the R_{pp} value was 0.96. This value indicates that 96% of the paper and plastic in the incoming waste was recovered in the light fraction. If the R_a of the lights had been 60, then 60% of the total ash of the raw refuse would have been carried over with the light fraction. (Actually, no tests were made of the lights to determine the R_a value.)

In terms of analysis, retained ash is a poor choice for a characteristic parameter, due to the fact that even a waste containing solely paper and plastic and which underwent air classification would have a retained ash value of 100%. However, the use of retained ash is a convenient one because ash data are available from a number of resource recovery plants for both shredded and air-classified refuse. Consequently, from a practical standpoint, retained ash is a convenient descriptive term. In addition, because the ash content of the inerts (inorganic materials) in the light fraction is commonly four to ten times that of the organic materials (mostly paper and plastic) in light fraction, the use of retained ash has a significance, albeit one that requires a knowledge of its limitations.

It is recommended by the authors that in the future, retained fines be used as a descriptive parameter in order to circumvent the aforementioned problems with retained ash. Retained fines could be designated as that material which passes a small screen mesh (perhaps 0.5 cm or less). Such material typically is 70 to 90% inorganic grit (mainly glass fines and dirt), and has a very high ash content, viz., on the order of 70 to 90%.

A third parameter, which could be termed the recovery index (θ), can be defined as $\theta \equiv R_{pp}/R_a$. The typical range of θ would be $0.9 \leq \theta \leq 2.5$. High values of θ are desirable and are indicative of a high percent recovery of combustible materials and a low percent recovery of materials with high ash content in the light fraction.

The following is a calculation that pertains to a practical operation, namely the St. Louis demonstration project.[6] Average results obtained in the demonstration indicate that the percent recovered paper and plastic and retained ash were respectively 0.93 and 0.73. Using the definition given in the preceding paragraph, the recovery index of the St. Louis air classifier was 1.3. Figure 6 shows schematically the derivation of the quantities mentioned above.

Although strictly speaking, paper and plastic are not the only materials that contribute to heating value of the light fraction, three reasons combine to make paper and plastic content a convenient indicator of yield of combustibles: (1) they supply the greater part of the heating value; (2) they can be rather easily identified and sorted; and (3) a data base exists in the literature for their measurement. Similarly, relative ease of testing and measurement prompted the choice of ash content as the measure of light fraction contamination. However, as previously noted, the use of measurement of inorganic fines in the waste streams would provide a better method of characterizing the contamination of the light fraction.

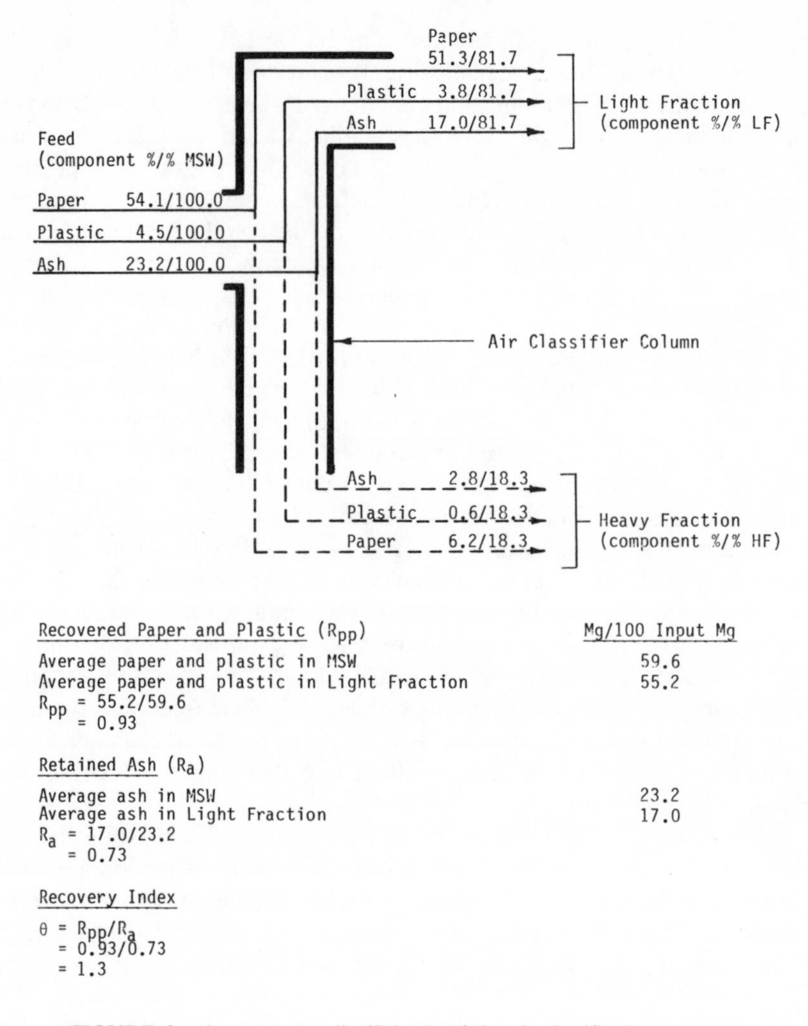

FIGURE 6. Average overall efficiency of the air classification process.

I. Air Classifier Performance

Values of percent recovery of paper and plastic (R_{pp}), retained ash (R_a), and overall recovery index (θ) for air classification can be determined for different operating conditions. Correlation of the performance parameters R_{pp}, R_a, and θ with operating conditions establishes the optimum conditions of α_{min}, β_{min}, V_o, q_o, and $\dot{m}_s$ for each air classifier. Through the use of information thus obtained, air classifiers can be compared among themselves as to separation performance, and extrapolations can be made regarding the effectiveness of specific design configurations.

An example of the latter application pertains to the utility of the baffles ("zig-zags") in the classifier column. Some authorities are of the opinion that the inclusion of baffles promotes agitation and mixing of the introduced material and thereby enhances separation.[9] The reasoning is that agitation and mixing in the zone of separation in the classifier control to some degree the separation of fine materials (e.g., glass and dirt fines) from components characterized by a large surface area and thin cross-section, as are paper and plastic. However, the reality does not necessarily coincide with the rationale. If the refuse is moist, fine materials tend to adhere to the larger surfaces. Moreover, fine particles have been embedded into large particles. Additionally, the aerodynamics of air classification are such that some dirt and glass particles inevitably are carried over with the light fraction regardless of hopefully preventive measures designed into the interior of the air column.

Judging from visual observation, it seems that material separation occurs mostly in the immediate vicinity of the feed opening of the 'classifier. Consequently, more rigorously attained test data than presently are available must be produced before it can be stated with any degree of certainty that the introduction of a feature to promote agitation is attended by an enhancement of separation efficiency.

J. Light Fraction Quality

Whereas percent recovered paper and plastic, retained ash or fines, and recovery index can be used to describe the air classification process, the quality of the light fraction (Q_{LF}) characterizes the air-classified product (light fraction). A possibly appropriate definition for the light fraction quality might be the following: "the weight ratio of paper and plastic to the total amount of light fraction expressed as

$$Q_{LF} = \frac{W_{Paper} + W_{Plastic}}{W_{LF}} \tag{13}$$

where Q_{LF} is the light fraction quality; W_{paper}, the weight of paper; $W_{Plastic}$, the weight of plastic; and W_{LF}, the weight of light fraction." This definition of quality differs somewhat from one proposed by Saheli.[10] The advantage of this definition over Saheli's rests in the ease of the testing and of the measurement involved in acquiring necessary data.

Typical values of light fraction quality are available from two reports in the literature.[6,11] One of the reports[6] deals with the St. Louis facility. According to data taken from it, the average lights quality as defined by Equation 12 was 68% on an as-received basis. Data from the second report, which was based upon University of California studies,[11] indicate an average light fraction quality of 66% on an as-received basis.

K. Optimization of Column Velocity

1. Definition of Column Velocity

Column velocity is another very useful parameter for describing the performance and operation of an air classifier. Characteristically, a zone of separation generally can be considered to extend from the opening through which the material is introduced to the bottom of the air classifier. Below the feed opening and well upstream of the heavies discharge, a characteristic average velocity can be described. This velocity is termed the "column velocity". Average column velocities can be determined experimentally through the use of pitot tubes and fluid manometers.[7] With respect to the classification of shredded solid waste, typical column velocities tried or in use range from 5 to 17 m/s.

If for some reason measurements of velocity cannot be made, an average column velocity can be estimated on the basis of the volumetric flow rate of air through the zone of separation and the cross-sectional area. The average column velocity is calculated by dividing the volumetric air flow rate by the cross-sectional area.

Usually, adjustment of column velocities is used as a means of regulating air classifier split. For example, if the velocity is increased above a nominal value, the percentage as light fraction tends to increase. When a wetter than normal waste is encountered, the column velocity usually is increased so as to maintain a constant split of material. Average values of column velocities at the Ames and the St. Louis facilities are approximately 17 and 11 m/s, respectively (see Table 4).

As of this writing, a rational basis for determining optimum column velocities apparently is yet to be developed and reported. Accordingly, visual observation of light and heavy fractions is the usual method currently employed to approximate a correct column velocity. For example, the appearance of paper and film plastic in the heavies or of aluminum contamination in the lights would serve as evidence of an unsuitable velocity. Since such a qualitative approach leaves much to be desired, the authors have undertaken the development of a suitable quantitative method. The development and the groundwork involved in the development form the subject matter of the sections which follow.

L. Aerodynamics of Air Classification

Given the highly experimental nature of air classification of shredded waste up to this time, a rational means must be developed for predicting and optimizing the performance of a classifier of a proposed design without being compelled to first build and test a working model that embodies the design. An appropriate rationale also would permit the assessment of the probable performance of an existing air classifier, given basic operating parameters of the unit such as volumetric air flow rate, solids flow rate, and cross-sectional area of the column.

One of the important fundamental concepts to be considered in the development of a rationale for optimization and assessment of performance is the Reynolds number. The Reynolds number is a dimensionless ratio of inertial forces to viscous forces within the fluid medium. Here, it owes its significance to the fact that air classification is a process which involves fluid flow (i.e., air flow) in a column. The importance of the Reynolds number is indicated by its inclusion in numerous relations concerned with fluid behavior and fluid influence upon submerged particles or bodies. Because of its unique and nondimensional formulation, the Reynolds number can be applied in the comparison and modeling of like phenomena of fluid dynamics, such as would be the case with different sizes of air classifiers.

1. Column Reynolds Number

In order to extend and generalize the theory of air classification to virtually all sizes and shapes, it first is necessary to establish the Reynolds number of the air flows within air classifier columns. In this section, the Reynolds number is calculated with respect to generalized classifier geometry, and is shown to indicate turbulent air flow in all cases where the flow is fully developed. Classifiers that are operated in the turbulent condition (i.e., Reynolds number > 3000) exhibit a comparable performance pattern that is based upon the principles of aerodynamic similitude.

The Reynolds number for fluid flow in a column is defined as

$$\text{Reynolds Number (R)} \equiv \frac{V_o D_h}{\nu_o} \quad (14)$$

where V_o is the free stream velocity; ν_o, the kinematic viscosity of air; and D_h, the hydraulic radius expressed as

$$D_h = \frac{4A}{P} \quad (15)$$

where A is the cross-sectional area and P, the perimeter.

For any arbitrary rectangular geometry of depth r and width s

$$A = rs \quad (16)$$

$$P = 2(r + s) \quad (17)$$

(s is perpendicular to material flow)

Substitution of Equations 16 and 17 into Equation 15 yields

$$D_h = \frac{2rs}{(r + s)} \tag{18}$$

Due to material distribution problems within the column, there generally is a practical range of ratios of width to depth. The ratio can be expressed as

$$K = \frac{s}{r} \tag{19}$$

and K generally assumes values between 1.0 and 6.0

It is convenient to express the hydraulic radius in terms of the size ratio K and the classifier width b by substituting Equation 19 into Equation 18:

$$D_h = \frac{2s}{(1 + K)} \tag{20}$$

Consequently, the Reynolds number can be expressed in terms of s and K by substituting Equation 20 into Equation 14:

$$R = \frac{2s}{(1 + K)} \frac{V_o}{\nu_o} \tag{21}$$

Values of the quantity $\frac{2}{(1 + K)}$ are within the range of 1.00 to 0.29 for $1.0 \leq K \leq 6.0$. Consequently, the interval of Reynolds numbers of interest is expressed as

$$\frac{0.29sV_o}{\nu_o} \leq R \leq \frac{1.00sV_o}{\nu_o} \tag{22}$$

In practice, average free stream velocities used in air classification fall within a certain practical range:

$$5 \frac{m}{s} \leq V_o \leq 17 \frac{m}{s} \tag{23}$$

With the air stream at 20°C, the kinematic viscosity (υ_o) is 1.49×10^{-5} m²/s. Consequently, the range of Reynolds numbers can be represented as

$$1.95 \times 10^4 \, sV_o \leq R \leq 6.71 \times 10^4 sV_o \tag{24}$$

where s and V_o are respectively in meters and meters per second.

Within the velocity range given by Equation 23, the range of Reynolds numbers becomes

$$9.74 \times 10^4 s \leq R \leq 1.14 \times 10^6 s \tag{25}$$

Because of the practical restraints on the width of an air classifier arising from the physical limitations on the width of in-feed conveyors or on rotary airlock feeders, the size range is

$$0.3 \text{ m} \leq s \leq 3.0 \text{ m} \tag{26}$$

Substituting Equation 26 into Equation 25 gives the practical range of Reynolds numbers encountered in air classification. The range is

$$2.9 \times 10^4 \leq R \leq 3.4 \times 10^6 \quad (27)$$

Since the Reynolds numbers exceed 3000 for those conditions specified by Equations 23 and 26, the flow, when fully developed within an air classifier column, can be considered turbulent throughout the sizes of air classifiers presently in existence or envisioned for the future. In actual circumstances the air flow within an air classifier is seldom fully developed due to the lack of a sufficient entry length at the bottom of the air classifier. The entry length phenomenon is addressed in the following section.

2. Entry Length Considerations

The entry length is the distance to the point of a fully developed turbulent velocity profile, and it is approximately 50 times the hydraulic diameter.[7,12] Consequently, a fully developed turbulent flow would be found in air classifier columns that are very long and have a small cross-sectional area. Thus far, air classifiers generally used experimentally and commercially do not conform to such a geometry. In fact, most tend to be roughly cubical in shape.

In the entry region, the air stream is accelerated from rest to its fully developed velocity profile. Due to the acceleration of the flow at entry locations, the Reynolds number increases from near zero, since the average velocity V_o is essentially zero at the entry, to numbers within the range $3 \times 10^4 \leq R \leq 3 \times 10^6$.

Since the greater number of air classifiers are not long enough to permit the establishment of a fully developed turbulent velocity profile, the description of the air movement becomes an entry length problem not amenable to analytical solution. Consequently, the establishment of the true velocity profile in air classifiers must be accomplished by actual velocity traverses within the classifier column. Inasmuch as the velocity profile is a developing one, a theoretical development of the profile would seem to be impossible to achieve at present because of complexities in the form of entry effects, and of the three-dimensional nature of the problem.

3. Drag Coefficients and Reynolds Numbers

The drag coefficient, C_d, is conveniently a function of the dimensionless Reynolds number, which is based upon a characteristic dimension of the object or particle. The relation is illustrated by the curves in Figure 7, in which the drag coefficient C_d is plotted as a function of the Reynolds number for circular cylinders, flat plates, and spheres, respectively. With respect to cylinders and spheres, the Reynolds number usually is based upon the diameter of the object. On the other hand, the Reynolds number for a flat plate is calculated on basis of the length of one side of the plate. Although there is no experimental evidence one way or the other, it may be assumed that the shape of the curve showing drag coefficients for cube-shaped particles should be somewhere between that of the curve for cylinders and that for spheres.

A range of Reynolds numbers can be calculated for the dimensions of particle sizes and column velocities commonly encountered in air classifying shredded refuse. The numbers are plotted in Figure 8. At sizes of spherical, cubical, and cylindrical particles from 1 mm to 5 cm, inclusively, and at column velocities from 6 to 17 m/s, Reynolds numbers that are based upon characteristic dimensions fall within the range of roughly 4×10^2 to 5×10^4. Within this range, the drag coefficients for the particles are approximately constant, as is indicated in Figure 7. With flat plates ranging from 0.3 to 20 cm and at column velocities of 6 to 17 m/s, the Reynolds numbers range from 3×10^3 to 2×10^5, and the drag coefficient is approximately constant.

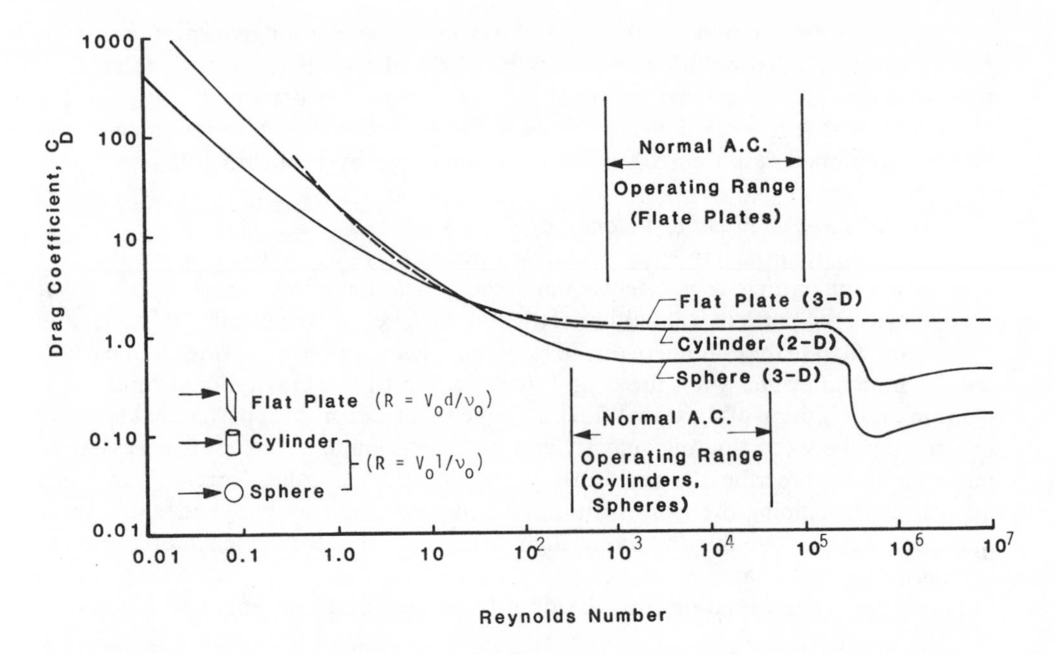

FIGURE 7. Drag coefficients for several geometries as a function of the Reynolds number.

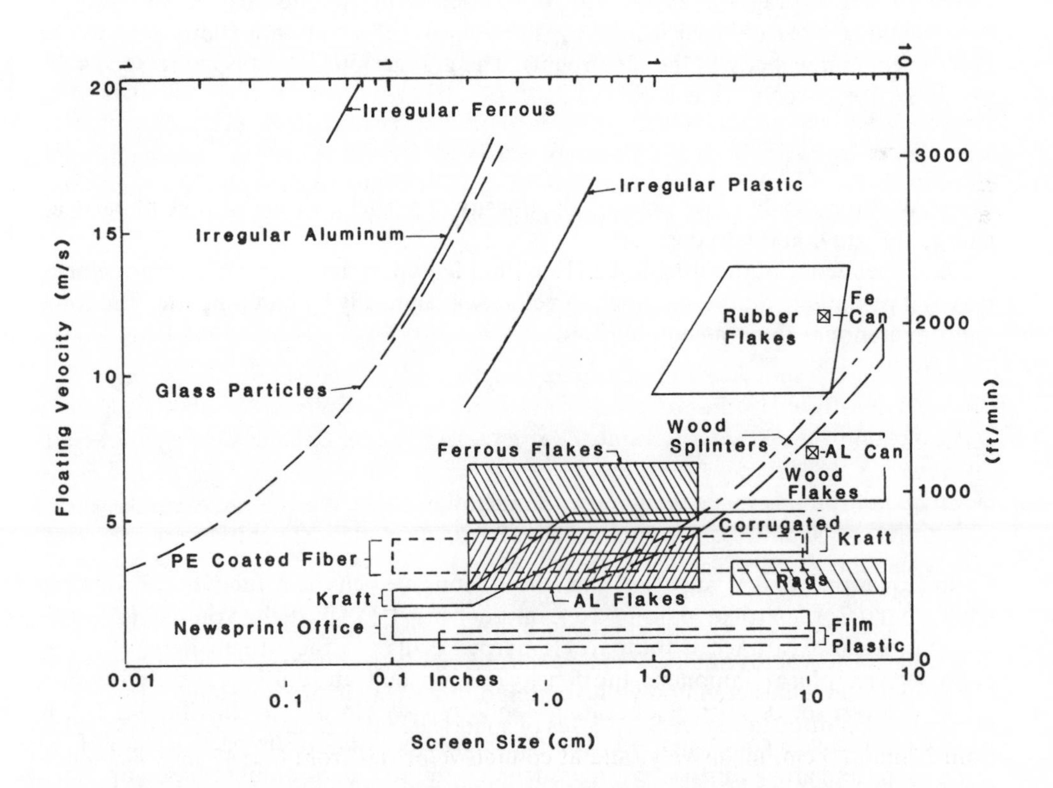

FIGURE 8. Floating velocities of shredded refuse components.

It should not be assumed that the values for drag coefficients for given particle shapes remain constant at Reynolds numbers less than 400 and at particle dimensions less than 0.4 to 1.0 mm. The lesser dimension applies at a column velocity of 17 m/s, and the larger at 6 m/s. A few relationships between drag coefficients for various shapes and Reynolds numbers less than 100 have been presented by Schlichting.[12]

4. Determination of Floating Velocity of Particles

In a column of upward moving air, an air velocity may be reached at which the drag force of a given particle is precisely equal to the gravitational force exerted upon it. At that velocity, the particle will neither settle nor rise (i.e., it is neutrally bouyant), and instead will float in the column of air. In brief then, in a column of moving air, a particle will be in equilibrium when the drag force is equal to the gravitational force.

From the definition of the term "floating velocity", it becomes apparent that to arrive at a relation between floating velocity and particle size and nature, it is necessary to mathematically describe the relationship between the opposing forces of drag and gravity. In developing the expression for floating velocity, the buoyant force can be neglected because of the relatively high density of the components of refuse as compared to that of air.

In developing the relationship, the drag force may be represented as follows:

$$D = \frac{1}{2} C_D \rho_o A_x V_o^2 \tag{28}$$

where D is the drag force; C_D, the drag coefficient; ρ_o, density of air; A_x, the cross-sectional area of particle; an V_o, the velocity of air stream (here used as the average column velocity of the air stream). The gravitational force is expressed as the weight of the particle, that is

$$W_s = \rho_s g Z \tag{29}$$

where W_s is the weight of particle; ρ_s, the density of particle; Z, the volume of particle; and g, the gravitational constant.

The dependence of the drag force (Equation 28) upon the square of the air velocity makes it possible to derive the floating velocity analytically by equating the drag force and gravitational force at equilibrium,

$$D = W_s \tag{30}$$

$$\frac{1}{2} C_D \rho_o A_x V_o^2 = \rho_s g Z \tag{31}$$

$$V_o = \left[\frac{2 \rho_s g Z}{C_D \rho_o A_x} \right]^{1/2} \tag{32}$$

From Equation 32, it is apparent that the floating velocity is a function of the drag coefficient (C_D), particle density (ρ_s), air density (ρ_o), particle volume (Z), and cross-sectional area (A_x) of the particle perpendicular to the stream flow.

For a given refuse component, the density ρ_s is constant. In addition, as stated earlier, under a given set of normal operating conditions within an air classifier, the drag coefficient for each component may be considered constant if the particles are of the same shape and are within a size range such that the Reynolds number based upon the characteristic particle dimension is within the range of 400 to 200,000. Assuming constant air temperature and pressure, the air density (ρ_o) also is constant. Therefore, the relationship for the floating velocity may be reduced to

$$V_o = \left[K \frac{Z}{A_x}\right]^{1/2} \tag{33}$$

where

$$K = \left[\frac{2g\rho_s}{C_D\rho_o}\right]^{1/2} \tag{34}$$

For a given material and set of air classifier operating conditions, the floating velocity (Equation 33) is a function of the square root of the ratio of the volume of the particle and its cross-sectional area perpendicular to the air flow.

A dimensional analysis shows that the floating velocity is proportional to the square root of the characteristic particle dimension. This relationship can be demonstrated by assuming a hypothetical particle of dimension L, and therefore volume L^3, and a cross-sectional area of L^2. The ratio of volume to area is directly proportional to L. Hence, the floating velocity is proportional to $L^{1/2}$. The $L^{1/2}$ dependence of the floating velocity implies that as a particle becomes smaller, the floating velocity decreases by the square root of the dimension L.

M. Aerodynamic Aspects of Particle Shapes

The actual process of air classification is a very complex one due to the broad collection of types of materials converted by shredding into myriad sizes and shapes. To account for these differences, it is necessary to consider the particle characteristics and the particle/air stream interaction in order to arrive at the values for drag coefficients required for the development of relationships for the floating velocities of the individual materials. The reason for the latter is the fact that the values of drag coefficients of particles are a function of particle shape, orientation of the particles in the air stream, and of a Reynolds number based upon a characteristic particle dimension.

Although it would be an exercise in futility to attempt to develop a precise and accurate analytical relationship for a drag coefficient for every conceivable particle shape, orientation in the air stream, and Reynolds number, several simplifying but valid assumptions can be made by means of which the task can be reduced to a manageable level. The assumptions stem from the particularities and a coupling of the shredding and air classification processes.

The task may be begun by considering the air classification aspects first. Earlier it was stated that in typical operations, the range of column air velocities is from 6 to 17 m/s. Air velocities within such a range, together with the typical size distributions resulting from shredding with the hammermill grate spacings at about 5 to 10 cm, combine to make the Reynolds numbers applicable to the particles to be within a range in which values of the drag coefficients are approximately constant for particular particle shapes. This fact justifies an assumption of invariance as to the values of the drag coefficients, provided of course, the particle shape remains constant. In summary, if the geometrical shape of a particular waste component remains unchanged throughout that component's size distribution, then it would be safe to assume that the drag coefficient value for the component also would be constant.

With respect to the shredding aspect, an important fact is that with certain components, the tendency is for the shape of the particles to remain relatively constant throughout the size distribution of the particles. For example, shredded particles of corrugated, newsprint, and ledger tend to assume the form of a flake throughout their entire size distribution. (A flake is characterized by a large area-to-thickness ratio.)

Particles formed by shredding ferrous and aluminum items are either in the form of flakes or of an irregular shape, as for example, a balled configuration. In the latter case, the form is dependent upon the grate spacing of the shredder. Particles formed in the size reduction of components other than the five named in the preceding sentences assume shapes that fall within four main categories, namely, flake, splinter, cylinder, and irregular. The invariance in particle shape combined with the range of Reynolds numbers characteristic of air classification permits the simplifying assumption that the drag coefficient is constant for specific components of shredded refuse.

To estimate the drag coefficient for each particle designation, it is necessary that an aerodynamic model be developed for each designation. Experimental data are available or can be inferred for flat plates, cylinders, and cubes.[12,13] The particle designations of flakes, cylinders, and irregular forms can be aerodynamically modeled respectively as flat plates, cylinders, and cubes. The modeling is summarized in Table 5.

Certain particle characteristics that are consequences of size reduction can be assumed for flakes, splinters, and irregular particles. Briefly these characteristics are (1) flakes—particle thickness much less than the characteristic particle length; (2) splinters (wood)—average length 15 times that of the diameter; and (3) irregular—particles that are roughly cubical in shape. (See items 1, 2, and 4 in Table 5.) Although shredding generally reduces beverage cans to noncylindrical particles, for the purposes of illustration the designation "cylinders" in this section refers specifically to whole metal beverage containers, the length of which is approximately 1.9 times the diameter (see item 3 in Table 5).

Drag coefficients for the particle shapes and characteristics generally encountered in shredded refuse have been estimated by the authors by adapting data taken from Hoerner[13] to information on the particle characteristics described in the preceding paragraphs. The estimated coefficients are 1.0 for flake type particles, 0.9 for splinters, 0.7 for cylindrical particles, and 0.8 for irregular particles (see Table 5). The coefficients necessarily represent average values, since the orientation of the particles in the air stream shifts continuously. It should be noted that because the values as listed for the coefficients take into account the geometry of the particles and the three-dimensional nature of the process, they differ somewhat from the idealized cases represented in Figure 7.

Using the drag coefficients listed in the preceding paragraph and assuming that the particles orientate themselves in the air stream such that they expose their maximum cross-sectional area to the direction air stream, it is possible to calculate the drag force. The soundness of the supposition regarding exposure of area was verified visually in experiments and by comparison with predicted results.

Further evidence in favor of the supposition can be found in the literature.[7]

The drag forces, $1/2\,(\rho_o C_D A_x V_o^2)$, as based upon information available at this juncture and expressed in terms of the longest particle dimension (L), are listed in Table 6. If the drag force for each particle designation is equated to the particle weight, then the floating velocity can be calculated by way of Equations 30, 31, and 32. A summary of the drag forces, weights, and floating velocities for each particle designation is given in Table 7. The floating velocities for splinters, cylinders, and irregular particles are a function of the square root of the particle length (L), whereas the floating velocities of flakes are a function of the square root of the particle thickness (t).

N. Determination of Floating Velocities of Shredded Refuse Components

Having determined the general relationships for the floating velocities characteristic of the various particle shapes, specific velocity relationships for shredded components of refuse can be developed. The development presupposes two pieces of information

Table 5
DRAG FORCES ON TYPICAL PARTICLE SHAPES

Particle shape	Particle characteristics	Particle designation	Aerodynamic model	Drag coefficient (C_D)	Maximum frontal area (A)	Drag force $\frac{1}{2}(\rho_o C_D A V_o^2)$
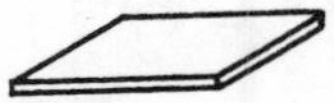	$t << L$	Flake	Flat plate	1.0	L^2	$0.5\rho_o L^2 V_o^2$
	$L = 15d$	Splinter	Cylinder	0.9	$Ld = 0.067L^2$	$0.033\rho_o L^2 V_o^2$
	$L = 1.9d$	Cylinder	Cylinder	0.7	$Ld = 0.53L^2$	$0.26\rho_o L^2 V_o^2$
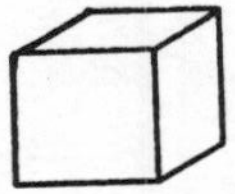	All sides of length L	Irregular particle	Cube	0.8	$L^2\sqrt{2}$	$0.57\rho_o L^2 V_o^2$

Table 6
DRAG FORCES OF THE FOUR CATEGORIES OF FORMS

Form	Drag force
Flake	$0.50\rho_o L^2 V_o^2$
Splinter	$0.033\rho_o L^2 V_o^2$
Cylinder	$0.26\rho_o L^2 V_o^2$
Irregular particles	$0.57\rho_o L^2 V_o^2$

Table 7
FLOATING VELOCITY RELATIONSHIPS FOR DIFFERENT PARTICLE SHAPES

Particle designation	Drag force (D)	Weight (W_s)	Floating velocity (V_o)	V_o(m/s)[a]
Flake	$0.50\rho_o L^2 V_o^2$	$g\rho_s L^2 t$	$1.41\left[\frac{g\rho_s t}{\rho_o}\right]^{1/2}$	$0.127\ (\rho_s t)^{1/2}$
Splinter	$0.033\rho_o L^2 V_o^2$	$g\rho_s \frac{\pi d^2}{4}$	$1.325\left[\frac{g\rho_s L}{\rho_o}\right]^{1/2}$	$0.029\ (\rho_s L)^{1/2}$
Cylinder	$0.26\rho_o L^2 V_o^2$	$0.22g\rho_s L^2$	$0.85\left[\frac{g\rho_s L}{\rho_o}\right]^{1/2}$	$0.077\ (\rho_s L)^{1/2}$
Irregular particle	$0.57\rho_o L^2 V_o^2$	$g\rho_s L^3$	$1.32\left[\frac{g\rho_s L}{\rho_o}\right]^{1/2}$	$0.199\ (\rho_s L)^{1/2}$

[a] When the units of ρ_s and t are Kg/m^3 and mm, respectively.

regarding particle characteristics, namely, density (ρ_s) of the material and the particle shape category of each component (i.e., flake, splinter, cylindrical, irregular). Having stipulated the two characteristics, the floating velocity relationship becomes a function of particle thickness (t) or length (L), depending upon the particle designation. Thus, the relationships for the floating velocities listed in Table 7 can be reduced to a specific velocity relationship for each of the components. Knowing the appropriate relationship and the characteristic dimension of a component, it is possible to calculate the specific floating velocity of that component. This procedure was followed in the determination of the floating velocities listed in Table 8. The table also includes data on moisture content, density, and characteristic dimensions.

The relationships for floating velocities developed in the preceding section can be used to determine a range of floating velocities for each refuse component as a function of the screen size or longest dimension of the particles. The development is illustrated in Figure 8. The information in Figure 8 pertains to refuse and is applicable to shredded material within the nominal range of 0.1 to 20 cm.

The curves in Figure 8 graphically demonstrate that a unique velocity does not exist at which combustibles (e.g., paper and PVC plastic) can be completely separated from glass fines and aluminum flakes. The impossibility of a complete separation stems in part from the inclusion of corrugated fiber in the light fraction. Glass and aluminum can be excluded from the lights only through selective sizing. This latter can be done by generating only large particles of glass or balled aluminum in the shredding operation. When in the balled form, ferrous and aluminum (designated as irregular ferrous and aluminum in Figure 8) have considerably higher floating velocities than when in the

Table 8
CALCULATED FLOATING VELOCITY RELATIONSHIPS FOR VARIOUS COMPONENTS OF SHREDDED REFUSE[a]

Waste component	MC (%)	ρ_s (kg/m³)	Particle designation	Typical characteristic dimension (t or L) (mm)	Floating velocity (m/s) as a function of t or L in mm	Typical floating velocity (m/s)
Newsprint	10	560	Flake	8.9×10^{-2}	$3.0t^{1/2}$	0.9
	40	840	Flake	8.9×10^{-2}	$3.7t^{1/2}$	1.1
Ledger	10	758	Flake	1.0×10^{-1}	$3.5t^{1/2}$	1.1
	40	1138	Flake	1.0×10^{-1}	$4.3t^{1/2}$	1.3
Corrugated S.W.[b]	10	192	Flake	3.7×10^{0}	$1.8t^{1/2}$	3.5
	40	320	Flake	3.7×10^{0}	$2.3t^{1/2}$	4.4
Corrugated L.B.[c]	10	650	Flake	3.2×10^{-1}	$3.2t^{1/2}$	1.8
	40	974	Flake	3.2×10^{-1}	$3.9t^{1/2}$	2.2
PE coated	10	746	Flake	7.4×10^{-1}	$3.5t^{1/2}$	3.0
	30	1066	Flake	7.4×10^{-1}	$4.1t^{1/2}$	3.5
PVC film	3	1008	Flake	2.5×10^{-2}	$4.0t^{1/2}$	0.6
PE film	3	912	Flake	5.8×10^{-1}	$5.8t^{1/2}$	4.4
PE	3	912	Irregular	5.8×10^{0}—1.8×10^{1}	$3.6L^{1/2}$	8.7—15.3
Lumber	12	480	Splinter	1.3×10^{1}—2.0×10^{2}	$0.6L^{1/2}$	2.2—8.5
	30	603	Splinter	1.3×10^{1}—2.0×10^{2}	$0.7L^{1/2}$	2.5—9.9
Plywood	12	552	Flake	3.8×10^{0}	$3.0t^{1/2}$	5.9
Rag	5	242	Flake	1.3×10^{0}	$2.0t^{1/2}$	2.3
Rubber	3	1773	Irregular	1.3×10^{1}	$5.0L^{1/2}$	18.0
	3	1773	Flake	2.5×10^{0}—5.1×10^{0}	$5.3t^{1/2}$	8.4—12.0
Aluminum	0	2688	Flake	1.3×10^{-1}—4.1×10^{-1}	$6.6t^{1/2}$	2.4—4.6
	0	2688	Irregular	2.5×10^{0}—5.1×10^{1}	$6.2L^{1/2}$	9.8—44.2

Table 8 (Continued)
CALCULATED FLOATING VELOCITY RELATIONSHIPS FOR VARIOUS COMPONENTS OF SHREDDED REFUSE[a]

Waste component	MC (%)	ρ_s (kg/m³)	Particle designation	Typical characteristic dimension (t or L) (mm)	Floating velocity (m/s) as a function of t or L in mm	Typical floating velocity (m/s)
Ferrous	0	7840	Flake	1.3×10^{-1}— 2.8×10^{-1}	$11.2t^{1/2}$	4.0— 5.9
	0	7840	Irregular	2.5×10^{0}— 5.1×10^{1}	$10.5L^{1/2}$	16.6— 75.0
Glass	0	2400	Irregular	2.5×10^{-1}— 1.5×10^{1}	$5.8L^{1/2}$	2.9— 22.5
Ferrous can	0	144	Cylinder	1.2×10^{2}	$0.9L^{1/2}$	9.9
Aluminum can	0	58	Cylinder	1.2×10^{2}	$0.6L^{1/2}$	6.6

[a] Based on experimental work done by the authors.
[b] S. W. = single wall (corrugated sandwich).
[c] L. B. = linerboard (1-ply corrugated).

FIGURE 9. Photograph of interior of experimental vertical air classifier.

flake form. As such the two metals are more likely to be separated into the heavy fraction.

Floating velocities of aggregate and dirt approximate those characteristic of glass particles, inasmuch as the densities of all three are comparable.

According to Figure 8, at an average free stream velocity of approximately 4 to 6 m/s, theoretically most of the incoming combustibles, i.e., paper, wood splinters, rags, and plastic materials, should be carried with the light fraction. Practical experience shows, however, that the velocities must be 20 to 40% higher than the theoretical to attain sufficient yields of combustibles in the light fraction. The failure of the actual velocities to precisely coincide with the theoretical velocities stems from the statistical nature of the process and the simplifying assumptions used in developing the theoretical velocities.

The theory on which are based the calculations for floating velocity relationships was verified in experimental studies by the authors. A vertical air classifier having a capacity of 4 tons/hr was used in the studies. The classifier was a rectangular column 2.5 m high, 1.1 m wide, and 0.4 m deep. A photograph and diagrammatic sketch of the unit are shown in Figures 9 and 10, respectively. Column velocities in the unit could be varied by adjusting the rate of air flow through the column.

In the experiments, the authors determined average column velocities through velocity traverses in which was applied the equal area method along the left, right, and center of the column at the midpoint of the separation area (see Figure 10). Velocities thus determined served as a basis for comparison with theoretical velocities. The average column velocities were corrected to a pressure of 1×10^5 N/m^2 and a temperature of 20°C. Floating velocities were defined as those at which 50% by weight of the particles reported to both the light and the heavy fractions. The materials used in the tests are described by the data in Table 9, in which are also listed the measured and calculated velocities attained with the materials.

A comparison between the measured and the theoretically calculated floating velocities listed in Table 9 indicates the existence of a close agreement between the two. With one exception ($\pm 13\%$), errors in the form of discrepancy between the two values

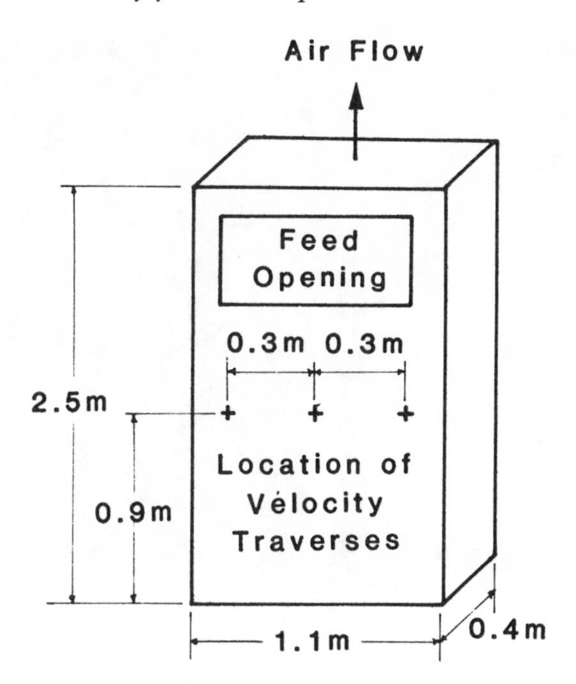

FIGURE 10. Diagrammatic sketch of experimental vertical air classifier.

were less than ±5%. Thus, despite the fact that the geometry of certain of the materials tested was not identical with that of the ideal models described in the preceding paragraphs, values based on the theoretical relationships for floating velocities were within an acceptable range.

An important consideration in the use of values calculated for floating air velocities is that the air classification process is necessarily a statistical process of separation. The reasons are twofold: (1) the turbulent nature and developing velocity profile of air stream within the column; and (2) the nonuniformity of the particles produced in the shredding operation. Thus, even with particles of a given material and shape that are within a narrow size range, some will fall and some will fly when the floating velocity is approached. This statistical nature accordingly compels the adoption of a definition of the floating velocity that implies a 50/50 split on a weight basis for each material, shape, and size range. The air velocity at which all particles of a given material, shape, and size will be propelled upward ("fly") is, of course, greater than at which they will be only buoyed ("float" and "floating velocity"). As with the floating velocity, "flying" velocity is contingent upon the nature, shape, and size of the material. It is not simply a percentage of the floating velocity.

In a separate set of experiments conducted by the authors, column velocities within the range of 6.6 to 7.6 m/s were shown to recover more than 95% of all input paper in the light fraction, providing the air dry moisture content of shredded raw waste was 33% or less.

The authors established the predominant orientation of particles in an air stream by visually observing the movement of corrugated fiberboard and aluminum cans suspended in the stream by means of lightweight thread. The arrangement is indicated in Figure 11. They noted that for the greater part of the time, the objects were aligned broadside to the stream velocity, as is indicated by the position of the objects in the figure. This orientation prevailed regardless of the point of attachment of the thread on the object.

Table 9
COMPARISON OF MEASURED AND CALCULATED FLOATING VELOCITIES OF CERTAIN REFUSE COMPONENTS

Material	Description	Dimensions (cm)	Density (kg/m^3)	Measured velocity (m/s)	Calculated velocity (m/s)	Error[a] (%)
Corrugated[b]	Single wall	15.2 × 15.2 × 0.38	229	3.58	3.74	+4.5
	Single wall	7.6 × 7.6 × 0.38	229	3.66	3.74	+2.1
Newsboard[b]	Box board	10.2 × 10.2 × 0.07	596	2.99	3.05	+1.9
Wood[b]	Splinter	9.5 × 0.7 × 0.3	481	6.27	6.25	−0.3
Glass	Fines	0.20[c]	2403	7.32	8.30	+13.4
	Fines	0.09[c]	2403	5.30	5.56	+4.9
Aluminum can	Beverage can	6.4 diam × 12.1	58	6.52	6.38	−2.1

[a] Error percent = (measured velocity − calculated velocity)/measured velocity.
[b] Oven dry moisture content = 10%.
[c] Average screen size of fines.

FIGURE 11. Orientation of objects in the column air stream. View of the interior of the air classifier column, looking vertically upward. Test objects can be seen floating in the air stream. During the greater part of their sojourn in the column, the objects are oriented broadside to the air stream. Two pitot tubes for velocity determinations can be seen in the lower center of the photograph.

O. Analytical Development of Composition and Properties of Air Classifier Fractions

With use of the analytical expressions for floating velocities and data on the size distributions of individual components of shredded refuse, a prediction can be made of the composition, heating value, and ash content of the light and of the heavy fractions. Moreover, it also becomes possible to predict both the classifier mass balance and the properties of the stream splits. To arrive at such predictions it is necessary not only to have the analytical expressions for floating velocities, but also those for the component size distributions as a function of particle size. A generalized development of these expressions is done by determining the dependence of cumulative weight percentage passing (w) for a given material as a function of a characteristic dimension (x), that is,

$$w = f(x) \tag{35}$$

where f(x) denotes "function of x", for example, a Rosin-Rammler equation describing cumulative weight percent passing vs. screen size. (Alternatively, cumulative weight percentage retained may also be represented.) Similarly, an expression for the floating velocity (V) of the given material as a function of the same characteristic dimension (x) has the form,

$$V = g(x) \tag{36}$$

where g(x) denotes function of x. The relations for the floating velocities shown in Table 8 are in the form of Equation 34, where the characteristic dimension x may be taken as t or L in the table as a first order approximation.

Manipulation of Equations 35 and 36 provides two expressions for the characteristic size (x). Consequently, Equations 35 and 36 can be equated to one another in the following manner:

$$x = f^{-1}(w) \tag{37}$$

$$x = g^{-1}(V) \tag{38}$$

$$f^{-1}(w) = g^{-1}(V) \tag{39}$$

$$w = f(g^{-1}(V)) \tag{40}$$

where $f^{-1}(w)$ and $g^{-1}(V)$ are inverse functions corresponding to f and g, respectively. This approach yields the mass fraction of a given material as a function of velocity.

Using the mass fraction expressed as a function of velocity, various properties of light and heavy fractions can be determined as a function of column velocity. This is accomplished through the arithmetic addition of those effects on the process that originate from the nature (physical properties) of each individual refuse component. The determination of a particular property (P) can serve as a generalized example of such an approach. In the example, P is chosen to be the heating value of a composite material—in this case, the light fraction. Then, if it is supposed that the composite material consists of i material components and has a total mass (m), and each component has the property p_i (e.g., heating value of component i), the mathematical formulation would be

$$P = \sum_{i=1}^{n} m_i p_i \tag{41}$$

where P is the property of composite material; m_i, the mass fraction of the ith component; p_i, the property of the ith component; and n, the number of component materials.

Applying Equation 41 to a given light fraction composed of mass fractions of paper (m_1), plastic (m_2), wood (m_3), and inert fines (m_4) with their respective heating values of hv_1, hv_2, hv_3, and hv_4, the heating value of the light fraction (HV) can be expressed as

$$HV = hv_1(m_1) + hv_2(m_2) + hv_3(m_3) + hv_4(m_4) \tag{42}$$

Despite the fact that the light and heavy fractions are composed of numerous individual components, simplifying yet realistic assumptions usually can be made regarding the relative contributions of individual components to the property of the composite, for instance, the assumption that the heating value of glass is zero. Consequently, only a few of the individual components may need to be expressed analytically in terms of their mass fractions as functions of column velocity and subsequent use in Equation 41. Moreover, certain individual components may be virtually independent of velocity within the range of the velocities under consideration. A specific application of the preceding approach has been explored by Fan.[14]

P. Summary of Design Considerations for Vertical Classifiers

1. Typical Design and Performance Characteristics

Data indicate that air classifiers can be treated generically as a first approximation in the determination of their ranges of performance.[15] The specific energy requirement of an air classifier is a function of the design of the unit and of its ancillary components (e.g., cyclones, baghouses). As listed in Table 10, the typical range is from 1.0 to 11 kWh/Mg.

2. Fluidization

The extent of the fluidization of the solid particles is indicated by the percentage of incoming paper, plastics, and fines that report to the light fraction. Percentages attained

Table 10
SUMMARY OF OPERATING AND PERFORMANCE PARAMETERS OF SEVEN AIR CLASSIFIER SYSTEMS

Parameter	High	Low	Typical range
Paper and plastic in heavy fraction (%)	42.8	0.8	5—30
Light fraction composition (%)			
Ferrous metals	3.3	0.0	0.1—1.0
Nonferrous metals	1.5	0.0	0.2—1.0
Fines	37.5	4.0	15—30
Paper and plastic	87.8	22.9	55—80
Ash	34.3	5.8	10—35
Percent of component retained in light fraction			
Ferrous metals	32.1	0.0	2—20
Nonferrous metals	96.7	0.0	45—65
Fines	100.0	68.4	80—99
Paper and plastic	99.9	69.5	85—99
Ash	85.2	29.3	45—85
Recovered energy	99.9	65.2	73—99
Specific energy[a] (kWh/Mg)	15.0	0.9	1—11
Column loading ($Mg/m^2/hr$)	>46.0	2.0	5—40
Recovered PP/retained fines[b,c,d,e]	0.9	1.1	1.0
Recovered energy/retained ash[f]	2.3	1.1	1.2—2.0

[a] Excludes freewheeling energy.
[b] PP = paper and plastic.
[c] Recovered PP = air dry weight of PP recovered in light fraction divided by air dry weight of PP in the air classifier feed.
[d] Retained fines = air dry weight of −14 mesh inorganic fines retained in light fraction divided by air dry weight of −14 mesh inorganic fines in the air classifier feed.
[e] Retained ash = oven dry weight of ash retained in light fraction divided by oven dry weight of ash in the air classifier feed.
[f] Recovered energy = $(M_f)_{LTS}\,(HV)_{LTS}/HV_{FEED}$
where M_f = mass fraction on an oven dry basis and
HV = high heating value on an oven dry basis.

with air classifiers representative of the present state-of-the-art generally range from 85 to 99% (14-mesh fines), as is shown in Table 10. Satisfactory separation of the cellulosic materials and fines from shredded refuse has been achieved at column loadings of 5 to 40 $Mg/hr\text{-}m^2$ (see Table 10).

3. Air-to-Solids Ratio

An adequate air dilution of the shredded material within the air classifier is an important consideration, because without it the degree of separation of combustibles and noncombustibles drops to an unacceptable level. Moreover, without it, the ability to absorb transient loading peaks is lost. Field tests[15] show that dilution is at a satisfactory level if the air-to-solids ratio on a weight basis is greater than 2, i.e., the critical ratio. Therefore, an air classifier should be designed to provide an air-to-solids ratio of 2 or higher. Theoretically, the upper limit of the ratio is solely that imposed by the dimensions of the mean free particle path above which the distance between particles is sufficiently great to allow adequate fluidization and separation by the air stream.

Table 11
AIR CLASSIFIER COST ESTIMATES

Site	Nominal capacity	Cost ($)	Costs per Mg of nominal hourly capacity ($)
Akron, Ohio	2 lines, 54 Mg/hr each	700,000 (1977)	6416
Ames, Iowa	27 Mg/hr	182,000 (1975)	6673
Chicago, Ill.	2 lines, 36 Mg/hr each	400,000 (1976)	5500
New Orleans, La.	27 Mg/hr	233,000 (1977)	8543
Niagara, N.Y. (Hooker Chem)	3 lines, 63 Mg/hr each	1,200,000 (1978)	6285
St. Louis, Mo.	27 Mg/hr	115,000 (1972)	4216
Tacoma, Wash.	73 Mg/hr	452,000 (1978)	6215

4. Recommended Specifications

Because of the nature and composition typical of shredded MSW, the column loading factor should be within a range of 7 to 15 Mg/hr-m^2; the air-to-solids ratio, 2 to 8; and the suggested range of average column velocities within the zone of materials separation, from 366 to 610 m/min (1200 to 2000 fpm). The ranges of the preceding three parameters are representative of the state-of-the-art of air classification at the time of this writing. Within those ranges, the gross specific energy requirement of an air classifier system, exclusive of baghouse, should be from 1 to 11 kWh/Mg processed. The ranges of performance to be expected are listed in Table 10.

The design criteria and range of performance parameters described in the preceding paragraph also can serve as specifications for air classifier designs. As such, one should bear in mind that dilution of the shredded waste by the air stream within the classifier column has a direct effect upon the efficiency of material separation. Consequently, the overall performance of an air classifier apparently improves until a column loading factor of approximately 7 Mg/hr-m^2 is reached. There is no evidence that material separation is improved by reducing column loading to less than 5 Mg/hr-m^2, nor is air classifier performance improved by exceeding an air-to-solids ratio of 7.

IV. COST OF AIR CLASSIFICATION

There is a paucity of cost data on air classification systems, chiefly because only a few such systems have been manufactured and installed. In the Ames and St. Louis operations, both of which process an average of approximately 30 Mg/hr, the system costs for air classification were $183,000 (1975 dollars)[5] and $115,000 (1972 dollars),[6] respectively. Since in both cases the figures are for total costs, they also include costs of the infeed system (surge bins, conveyors, etc.) of the vertical air classifier and of the pneumatic system (ducting, blower, cyclones, etc.).

The total costs of several other air classification systems are shown in Table 11. These costs are modifications of those by Chrisman.[8] On the basis of nominal hourly capacity, the cost per ton as shown in Table 11 ranges from approximately $3800 to $7800.

Generally manufacturers of air classifiers insist upon controlling the entire air classification system, including the feed, air classifier, the pneumatic subsystems, to assure the satisfactory operation and performance of their equipment. Consequently, the equipment cost is a moot question since engineering costs are incorporated into the total system cost.

REFERENCES

1. Unpublished data, University of California (Berkeley), 1982.
2. **Murray, D.,** Air Classifier Performance and Operating Principles, presented at the 1978 Natl. Waste Process. Conf., Chicago, May 7 to 10, 1978.
3. **Murray, D.,** Personal communication, 1979.
4. **Murray, D. and Liddell, C. R.,** The dynamics, operation and evaluation of an air classifier, *Waste Age,* 8, 18, 1977.
5. **Even, J. C., Adams, S. K., Gheresus, P., Joensen, A. W., and Hall, J. L.,** Evaluation of the Ames Solid Waste Recovery System. I, U.S. EPA, Cincinnati, Ohio, 1977.
6. **Fiscus, D. E., Gorman, P. G., Schrag, M. P., and Shannon, L. J.,** St. Louis Demonstration Final Report, EPA-600/12-77-155a, U.S. EPA, Cincinnati, Ohio, September 1977.
7. **Jorgensen, R.,** Ed., *Fan Engineering,* Buffalo Forge Company, New York, 1970, 486.
8. **Chrisman, R. L.,** Air Classification in Resource Recovery, National Center for Resource Recovery, Inc., RM 78-1, October 1978.
9. **Boettcher, B. A.,** Air classification for reclamation of solid wastes, *Compost Sci.,* 2 (6), 22, 1970.
10. **Saheli, F. P.,** The technology of solid waste air classification for resource recovery, *A.S.M.E. 76-ENAs-50,* 1976, 7.
11. **Trezek, G. and Savage, G.,** MSW component size distributions obtained from the Cal resource recovery system, *Resour. Recovery Conserv.,* 2, 67, 1976.
12. **Schlichting, H.,** *Boundary-Layer Theory,* McGraw-Hill, New York, 1968, 560.
13. **Hoerner, S. F.,** *Resistance a L'avancement Dans Les Fluides,* Gauthier-Villars Editeur, Paris, 1965, 44.
14. **Fan, D.,** On the air classified light fraction of shredded municipal solid waste—composition and physical characteristics, *Resour. Recovery Conserv.,* 1 141, 1975.
15. **Savage, G. M., Diaz, L. F., and Trezek, G. J.,** Performance characteristics of air classifiers in resource recovery processing, in Proc. 1980 Natl. Waste Process. Conf., A.S.M.E., 1978.

Chapter 6

TROMMEL SCREENING

I. STATE OF THE ART AND FUNDAMENTALS

A. Utilization of Screens in Solid Waste Processing

Screens can be used in solid waste processing to achieve an efficient separation of refuse particles on the basis of differences in physical size in any two dimensions. The separation results in a division of the incoming material into two size groups in which one has a minimum particle size larger than that of the individual openings of the screen (or perforated plate), and the second, a maximum particle size smaller than that of the openings. The former are retained on the screen (oversized particles, "oversize"), whereas the latter pass through it (undersized particles, "undersize").

Although several types of screens are used in industry, only three have been used for sizing particular fractions of processed and unprocessed municipal solid waste. The three types are the vibratory flat bed screen, the disk screen, and the trommel screen. Experience with the flat bed and the disk screens has been only doubtfully successful. On the other hand, the trommel screen has been proven to be quite effective and efficient.

1. Trommel Screen

The trommel is a rotary cylindrical screen whose screening surface consists of a wire mesh or perforated plate, as is indicated by photograph in Figure 1. It can be used to process raw refuse prior to size reduction ("pretrommeling"*), and to process shredded refuse ("posttrommeling"). With either option, use is made of the characteristic tumbling action of the rotary screen to effect efficient separation, albeit for slightly different end results.

The trommel is an especially efficient device due to its inherent capability of tumbling the throughput material. The tumbling action brings about a separation of individual items or pieces of material that may be attached to each other, or even of one contained within another. The tumbling action is virtually essential if a material such as municipal solid waste is to be screened. The reason is the need for a high degree of screening efficiency coupled with a minimum of screening surface.

The size of the trommel screen openings can be manipulated to allow the concentration of certain refuse components either in the oversize fraction or in the undersize fraction, as the need may dictate. If the component size distributions of particular constituents of a mixture are known, as for example those shown for a hypothetical raw municipal solid waste in Figure 2, the effect of a given screen mesh upon the resultant size distribution of each component after screening can be ascertained. Consequently, with a prior knowledge of the size distribution of the components of a processed refuse, it is possible to design an efficient screening process.

The utility of a component size distribution in designing a screen for concentrating given materials is illustrated in Figure 3. According to the figure, the paper and plastic components in the sample analyzed in the figure were concentrated in the +10-cm material. As the figure shows, the percentage of paper and plastic materials peaked within the size range −20 to +10 cm. Consequently, in terms of paper and plastic

*Strictly speaking, the parenthetical terms "pre-" and "posttrommeling" in the paragraph are not accurate, because as used, the prefixes "pre" and "post" actually refer to shredding rather than to trommeling. However, because the terminology in the paragraph is widely used in the industry, it is so used in this book.

FIGURE 1. Photograph of trommel screen.

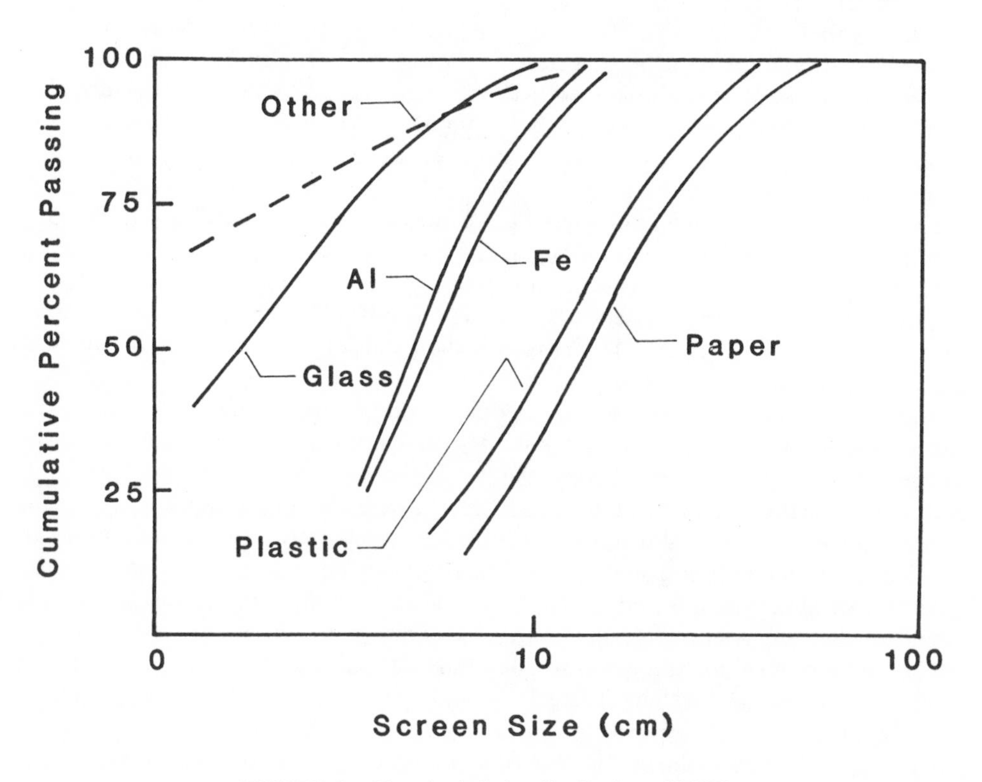

FIGURE 2. Hypothetical size distribution of MSW.

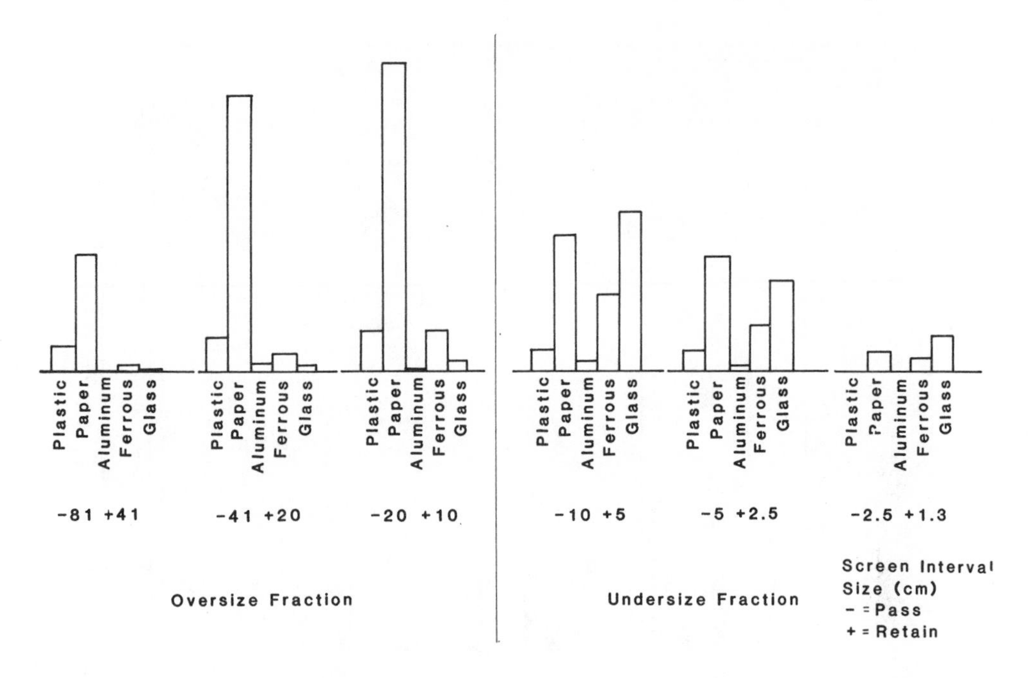

FIGURE 3. Component mass distribution of MSW.

recovery (or synonomously, refuse-derived fuel production), the trommel screen design should call for a 10-cm opening.

The method of analysis described for raw municipal solid waste in the preceding paragraph can also be applied to the air-classified light fraction. The data developed in making the analysis of the mass distribution of the components are plotted in Figure 4. In the sample of air-classified lights analyzed through the use of the two figures, the percentage of paper (designated "fiber" in the figures) and of plastic materials declined significantly when the screen size openings were less than 0.94 cm. Therefore, to obtain maximum recovery of paper, and hence a high quality RDF, the screen openings of the trommel should be 0.94 cm (0.375 in.).

2. Pretrommel Screening

The rationale for pretrommeling is based on the fact that not all components of refuse require size reduction prior to downstream processing. In the state in which they occur in raw refuse, certain components already are smaller than the size that would be imposed by size reduction. Furthermore, an extension of this reasoning leads to the generalization that unnecessary size reduction complicates downstream processing by multiplying the number of particles that must be separated. Despite the preceding and other potential advantages that accrue from pretrommeling, to date there has been a dearth of data reported in the literature describing the advantages.

The performance to be expected in a typical pretrommeling process may be illustrated by results obtained with refuse received at the Recovery-I plant in New Orleans.[1] According to a mass balance made of the operation of the plant's pretrommel unit, the split of material to the oversize fraction was approximately 43.5% and to the undersize fraction, 56.5%. In other words, the incoming raw waste was separated into two fractions roughly equal in size on a weight basis, one of which was oversized material, and the other undersized material. The greater part of the glass in the raw wastes was in the undersize fraction, while most of the combustibles (paper and plastic)

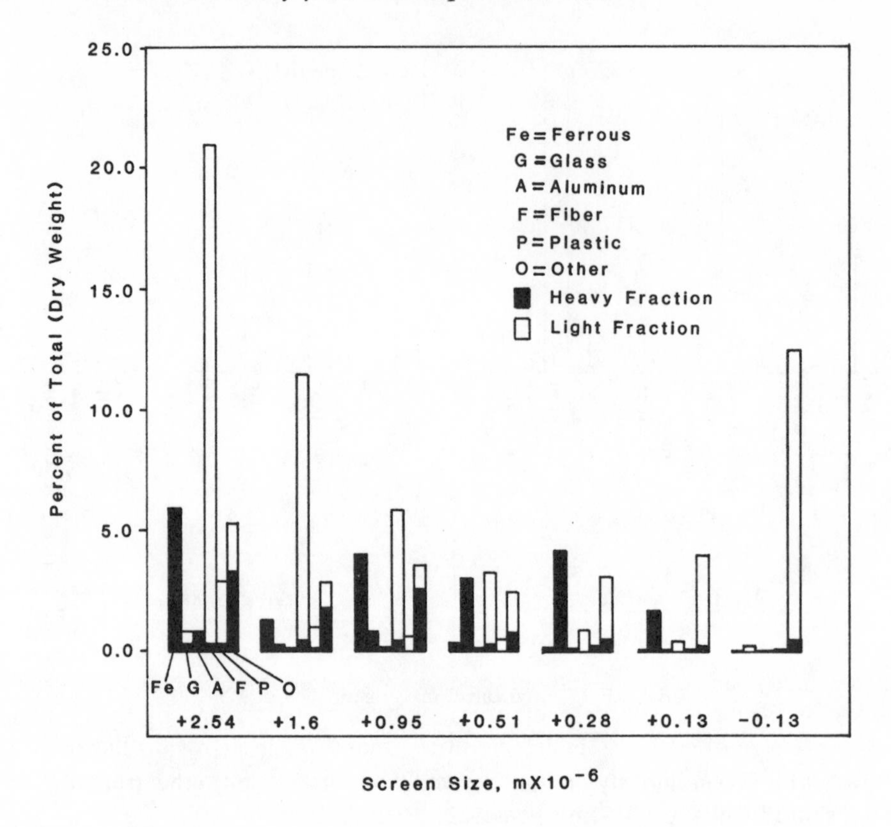

FIGURE 4. Component mass distribution of air-classified MSW.

was in the oversize fraction. However, the degree of the separation of the glass was more complete than that of the combustibles, in that approximately 99% of the glass was discharged in the undersize fraction. On the other hand, since only 67% of the incoming paper and only 60% of the plastic was found in the oversized fraction, a considerable amount of the paper and plastic was also to be found in the undersize material. Typically, the heating value of the oversize fraction was on the order of 16,002 kJ/kg (7150 Btu/lb) dry weight; the ash content, 10%; and the moisture content, 26%.

B. Trommel Screening the Air-Classified Light Fraction

1. Benefits

Trommel screening is an effective means of upgrading the air-classified lights (ACLF) because through it inorganic material is removed from the paper and plastic fractions. The quality of the ACLF is improved for use in two important applications, namely as a refuse-derived fuel (RDF) and as a mixed wastepaper feedstock.[2] The improvement in quality as an RDF is brought about by the removal of fines, which are largely inorganic or "inert" in nature and hence are noncombustible. As such they detract from the heating value of the unscreened lights and add to the amount of ash remaining after combustion. An added feature, probably only partially due to the nature of the fines, is a loss of moisture, and consequently a further increasing of the heat content of the screened lights.

The upgrading effect of trommel screening the lights is illustrated in a concrete manner by the data in Table 1, in which are presented the compositions of unscreened and of screened lights. The compositions given in the table are quite typical. If the data

Table 1
COMPOSITION OF AN AIR-CLASSIFIED LIGHT FRACTION AND ITS SCREENED LIGHT FRACTION

	Air-classified lights		
		After screening	
Component	Before Screening, weight (%)	"Oversize" weight (%)	"Undersize" weight (%)
Paper	60.2	78.8	19.9
Plastic	5.7	7.8	1.4
Miscellaneous	34.1	13.4	78.7
	100.0	100.0	100.0

are representative—and they probably are—then, on an as-processed basis, paper would be expected to comprise on the order of 60% and plastics about 6% of MSW air-classified lights. The remaining 34% would be a miscellany of materials, mostly fines. On the other hand, paper and plastics could constitute 86 to 87% of the screened lights, thus leaving only 13 to 14% in the form of fines.

Improvement of the air-classified lights as a wastepaper feedstock is to the extent that screened lights can be successfully utilized as a supplement for virgin pulp in pulp and paper manufacturing. The upgrading in quality is brought about by the removal of inorganic fines. When present in the feedstock, fines result in an overloading of the pulp cleaning equipment. Moreover, organic fines increase the biochemical oxygen demand (BOD) of the wastewater produced in the pulping process.

2. *Principles*

At the time of this writing, the majority of the work done on the trommel screening of the air-classified light fraction had been carried on at the University of California, Berkeley.[3,4]

When screening air-classified lights, the trommel screen openings should be sized such that the relatively large particles of paper and plastic are segregated from the finer materials that in the air classification step may have been carried over with the large particles into the receptacle for the light fraction. The fine materials consist mainly of dirt, metals, and glass. Their carry over in air classification results from their characteristically large drag-to-weight ratio. Trommel screens are specifically used for this separation because the tumbling motion of the particles imparted by the rotating screen surface in combination with the influence of gravity displaces any fine inorganic materials that may be adhering to or are stratified among the paper and plastic particles. Not unexpectedly, the oversize fraction of the screened lights is termed "screened light fraction", or simply "screened lights".

3. *Processing Comparisons*

As was explained in two preceding sections, pretrommeling and trommel screening the light fraction are two methods for increasing the energy content and lowering the moisture and ash content of refuse-derived fuel. Unfortunately, experimental data and reports of actual experience on which can be based hard and fast conclusions on the relative merits of the two processing alternatives with respect to each other are very few in number. However, with the sparse information that is available,[1,2,4] a rough comparison can be made between the two screening alternatives in terms of effect on quality of the refuse-derived fuel. Data for making such a comparison are listed in Table 2.

Table 2
COMPARISON OF PRETROMMELING AND LIGHT FRACTION SCREENING

	Pretrommel oversize	Screened light fraction
Mass yield (%)[a]	43.5	47.9
High heating value (kJ/kg)[b]	16,600	18,600
Moisture content (%)[b]	25.8	16.3
Ash content (%)[b]	9.6	12.1
Energy yield (kJ/kg MSW)[c]	7,200	8,900
Sulfur (%)[b]	0.1	0.1

[a] Mass yield (%) = kg of oversize/kg of MSW × 100.
[b] Oven dry weight basis.
[c] Energy field (kJ/kg of MSW) = mass yield × high heating value.

II. THEORETICAL ASPECTS OF PARTICLE MOTION WITHIN A TROMMEL SCREEN

A mathematical description of the motion of refuse particles within a trommel screen is necessarily an extremely complicated undertaking owing to the heterogeneity of the material and its concomitant deviation from an "ideal" material in the sense of classical particle dynamics. In addition to the problem of nonhomogeneity, it is necessary to contend with complications such as:

1. Three-dimensional particle trajectories
2. Timing the rate of change of mass of material at successive longitudinal positions in the trommel due to loss of undersize material through the screen openings (shown in Figure 5)
3. The unknown contribution of particle moment of inertia to the overall motion and trajectory of the mass of material
4. The unknown magnitude of aerodynamic forces acting upon the particles
5. The unknown magnitude of frictional forces existing between the screen surface and the refuse particles
6. The potential influence of particle-particle forces, such as friction and adhesion

Although simplifications of the theoretical aspects of the motion of particles within a trommel can be made in the hope of rendering a problem amenable to an analytical solution, the simplifications may lead to erroneous conclusions when the flow of material is examined in the field or in the laboratory. An example of such a simplification would be the excluding of items 2 through 6 of the preceding six listed items. Because of this possibility, results predicated upon a theoretical development should be verified experimentally. In fact, on the basis of preliminary experiments, the authors have come to the conclusion that results obtained through the application of the classical theory of dynamics to a description of particle movement within a trommel are accurate only within an order of magnitude. The difficulty is illustrated in the succeeding paragraph.

The general motion of the flow of material within a trommel is diagrammed in Figures 5, 6 (Cases I and II), and 7. For the purpose of the illustration, the diagram in Figure 6 does not take into account the flow of some particles through the screen openings. The particle motion diagrammed in Figure 7 reflects the governing equations of motion

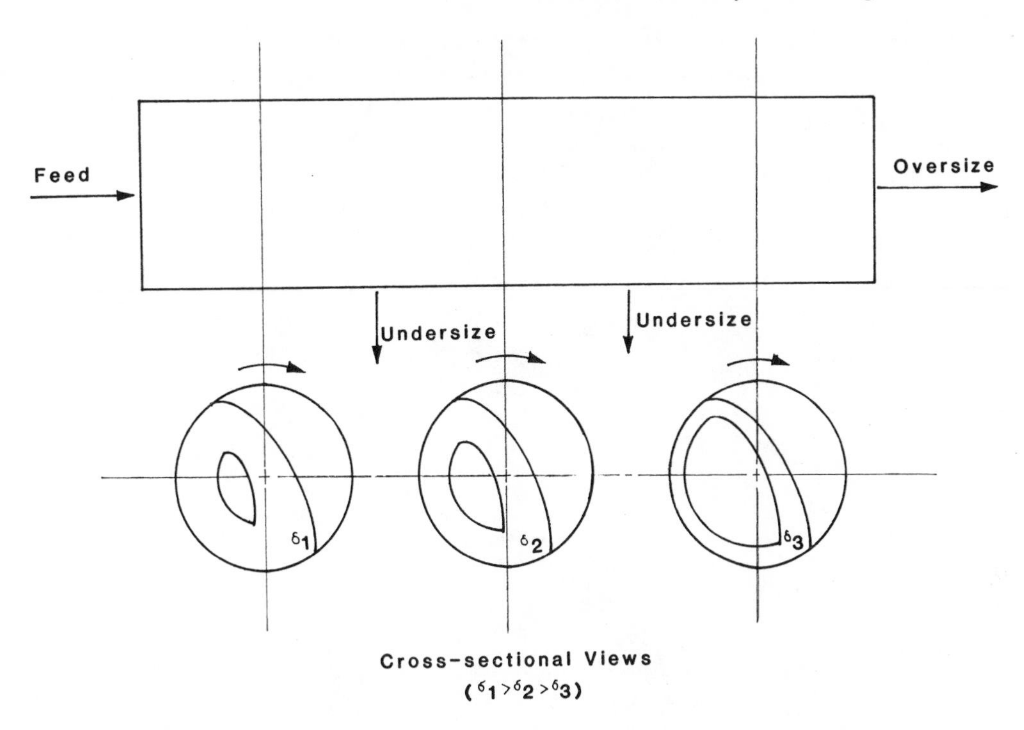

FIGURE 5. Variation in bed depth along longitudinal axis of a trommel screen.

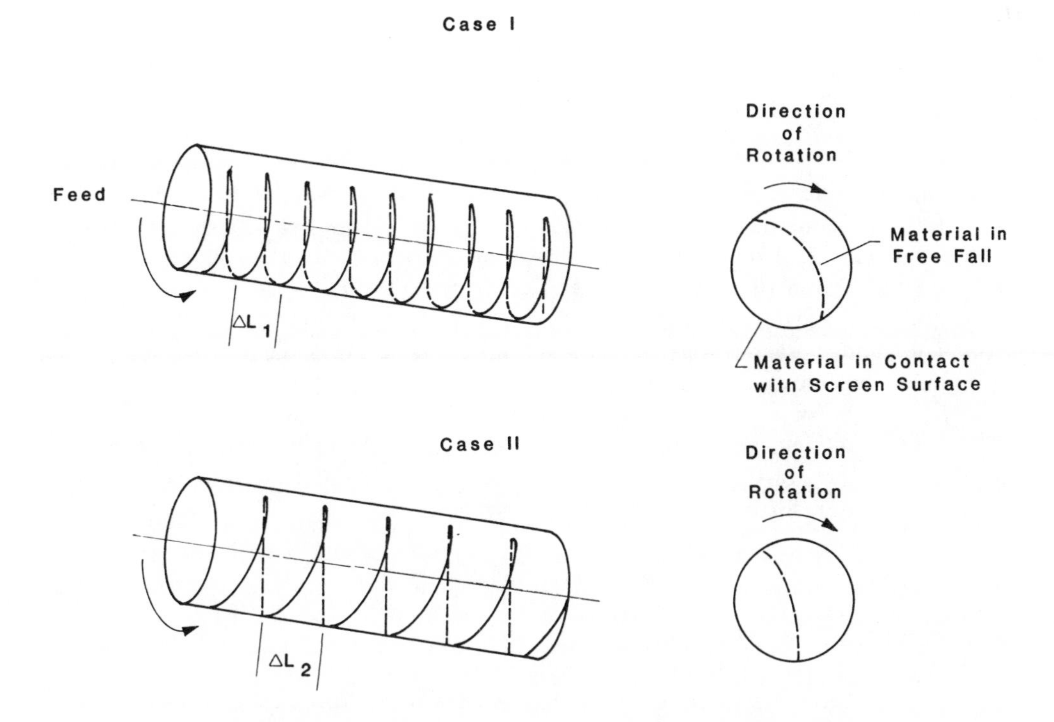

FIGURE 6. Flow pattern of material within a trommel screen.

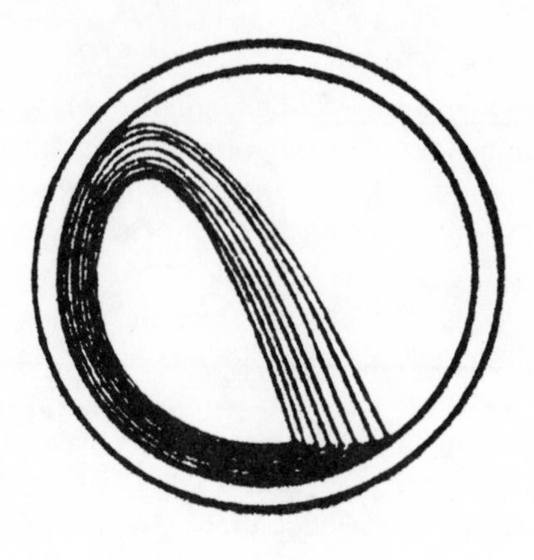

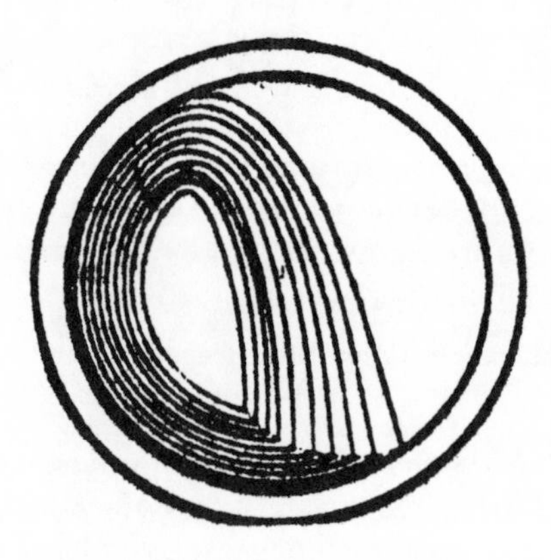

FIGURE 7. Computerized plot of movement of particles at various rotational velocities.

based upon classical dynamics and a computer simulation developed by the authors. In Case I, the incremental distance of travel of a single particle per screen revolution (ΔL_1) is less than that (ΔL_2) in Case II. Case I shows the classical particle trajectory. It is based only on the action of gravitational and centripetal forces. Case II represents the trajectory of particles as based on observations made in preliminary experiments conducted under practical operating conditions by the authors. Thus, according to the authors' observations, the reality is that particles tend to migrate through a trommel at a rate faster than that predicted by classical theory.

Discrepancies that arise from the neglect of items 2 through 6 can be lessened by adding the effects of the items in a systematic manner to the governing equations of motion and by resorting to analysis by computer.

III. PARAMETERS

A. Screening Efficiency

Regardless of type of screen, screening efficiency is an important parameter in a description of its performance. Efficiency generally is expressed as the amount of undersize material actually removed by the screen divided by the total amount of undersize material present in the feed material. The latter quantity is determined independently by passing a sample of the feed through a laboratory screen.

A highly efficient screen produces an oversize material that contains only a modicum of undersized material. In short, screening efficiency is a measure of the degree of separation of materials based solely on the criterion of particle size. The efficiency attained with a screen is principally a function of the physical properties of the input material, degree of agitation of the material while on the screening surface, depth of material on the surface of the screen, and residence time of the material within or on the screening device.

Materials that both tend to be granular in nature and to have a specific density greater than approximately 320 kg/m^3 can be screened with relative ease, in that relatively high efficiencies are attained at short residence times. On the other hand, a long residence time is needed to attain comparatively high efficiencies if the material is not strictly granular, and is instead, a composite of shredded paper, glass, and metals that has a broad range of sizes and a density less than 240 kg/m^3. Unfortunately, most of the materials destined to be processed in a resource recovery facility fall under the latter description.

The attainment of an effective sizing separation of a material stream requires a screen surface area large enough to provide an adequate residence time and an adequately thin bed depth of material. Therefore, the surface area necessarily becomes very large when shredded paper and plastics are components of the feed material. The throughput rate at which a 90% screening efficiency can be attained with a trommel processing the latter material is only about 0.10 Mg/m^2.

1. Residence Time

With a given material and depth of bed, screening efficiency is primarily a function of residence time, that is, the average length of time the material is within the trommel screen or on the screening surface. The longer the residence time, the higher becomes the screening efficiency. The relationship is illustrated by the curves in Figure 8. Because of the low density of refuse and its wide spectrum of particle sizes, a long residence time generally is needed to screen it at a satisfactory level of efficiency.

Residence time can be defined in terms of the length of the screening surface and the average velocity at which the material moves through (trommel) or over a screen (flat-bed), i.e., as

$$\tau = L/\bar{V} \tag{1}$$

where τ is the residence time, L is the length of screening surface, and $\bar{V}$, is the average longitudinal velocity of material through or over the screen. Inasmuch as for a given set of parameters of screen operation, the average velocity of material as it moves through or over the screen surface ($\bar{V}$) is fixed, the residence time can be prolonged by lengthening the screen (L).

Average velocity ($\bar{V}$) is determined by a number of variables. However, as of this writing, data on the identity and effect of the variables are almost completely absent because of a lack of in-depth studies on the subject.

Because of a dearth of data, at this time it is difficult to quantify the effect of residence time on screening efficiency. According to Woodruff,[5] a residence time on the order of 25 to 30 sec. is required for "good" separation of raw refuse in pretrommeling. His definition of "good" appears to be qualitative rather than quantitative, although his results do imply separation efficiencies of at least 75% for the residence times cited by him. Woodruff's times are at the short end of the range recommended by Hill,[6] namely, 30 to 60 sec.

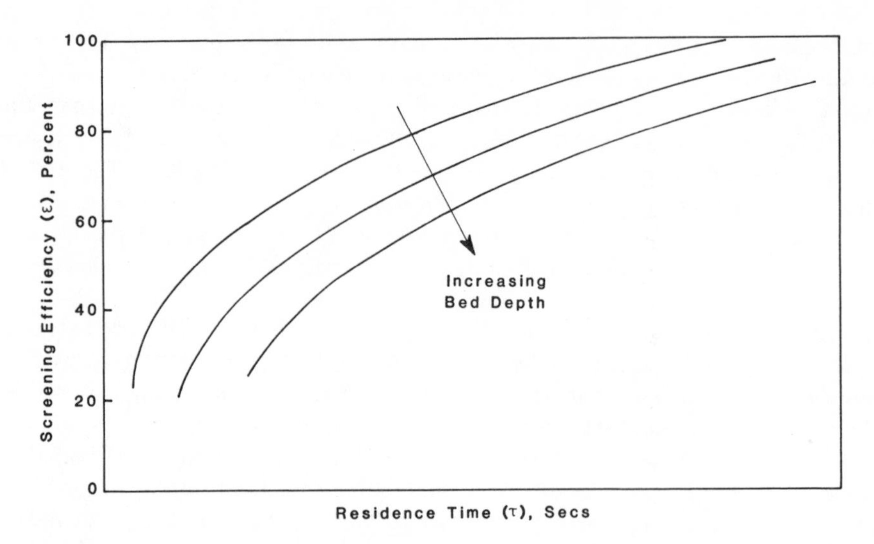

FIGURE 8. Screen efficiency as a function of residence time.

In unpublished research conducted at the University of California (Berkeley), it was found that separation efficiencies of 90% or better were obtained at a residence time of approximately 10 sec. when air-classified "lights" were processed in a trommel screen. The difference between the residence times quoted by Woodruff and Hill and that observed at the University of California is explained by the fact that air-classified "lights" and raw refuse are very different materials and that Woodruff and Hill included particle breakup time and particle separation time after breakup in their estimates. (Particle breakup refers to the opening of bags and containers.)

2. Bed Depth

The depth of material on the screen surface (termed the bed depth) determines the screening efficiency for a specified residence time. Consequently, to attain a high screening efficiency and yet have a short residence time consonant with a minimum of screening surface area, it is necessary to maintain the bed as a shallow layer. Obviously, the thinner the layer, the more readily and hence more likely it is that the fines will settle to the screen surface and, hence, through the screen openings.

Although bed depth is a straightforward quantity in flat bed screening, its definition in trommel screening is rather nebulous. The nebulosity is due in part to the fact that in trommel screening, the tumbling material has a distorted cross-section such as is indicated by the diagrams in Figures 5 and 6 and an ill-defined density. To arrive at a firm definition, a first step must be the characterization of the bed depth parameter to be used in trommel screen theory and in the development of the design guidelines.

3. Critical Frequency

A parameter that lends itself to engineering analysis in the trommel screening of solid waste is known as "critical frequency". It is expressed in units of revolutions per minute. The critical frequency of a trommel screen may be defined as that frequency of rotation at which material in it is held by centripetal force against the cylindrical screening surface throughout a complete revolution. The definition presupposes that no slippage of material occurs at the screen surface. At the critical frequency, the centripetal force is exactly balanced by the gravitational force at the apogee of the screen surface. At frequencies greater than the critical value, particles are not tumbled in the trommel nor do they axially traverse the cylindrical screen. Therefore, trommel screens are designed to operate at less than the critical value.

The governing relation between the diameter of the trommel and its critical frequency can be determined by making a simple force balance for a particle located on the cylindrical surface of a rotating trommel. The balance is one between two opposing forces, namely gravity and centripetal force. For the material to tumble and consequently move axially through the trommel, the individual particles must fall away from the screening surface before reaching the apogee. In theory the fall of the particles should occur immediately prior to reaching apogee so as to maintain a shallow bed of material on the screen and thereby raise the screening efficiency. Consequently, an ideal operation would entail a rotational velocity just short of the critical frequency. However, in the solid waste industry frequencies are usually in the range of 30 to 50% of the critical frequency, probably due to the mimicry of design criteria used in the mineral processing industry. To determine the proper point of fall, the relation between the diameter of the trommel and its critical frequency must be defined. The steps involved in reaching such a definition are defined in Equations 2 through 8.

The governing equation appropriate for zero displacement at the apogee of particle motion is

$$\Sigma F = 0 \tag{2}$$

Substituting the gravitational force (mg) and the centripetal force (mV^2/r), Equation 2 becomes

$$mV^2/r - mg = 0 \tag{3}$$

$$V^2 = rg \tag{4}$$

$$V = (rg)^{1/2} \tag{5}$$

where: m is the mass of the particle; r, the radius of trommel screen; a, the gravitational constant; and V, the tangential velocity at which the screen surface is rotating.

The critical frequency (f) of the trommel is derived from the relation

$$f = V/2\pi r \tag{6}$$

Solving Equations 5 and 6 for V and equating them allows development of a relation for f in terms of trommel radius (r),

$$f = \frac{1}{2\pi r}(rg)^{1/2} \tag{7}$$

or

$$f = \frac{1}{2\pi}\left[\frac{g}{r}\right]^{1/2} \tag{8}$$

In SI notation Equation 8 becomes

$$f = 29.9r^{-1/2}$$

where the units of f and r are revolutions per minute and meters, respectively.

Table 3
SCREEN CAPACITIES AND RELATED SEPARATION EFFICIENCIES

Type of trommel	Nominal throughput (Mg/hr)	Total screen surface area (m^2)	Screen capacity (Mg/hr-m^2)	Nominal separation efficiency (%)
Pretrommel	55	132	0.41	80
Light fraction	1.4	13	0.11	90
Light fraction[a]	1.8	13	0.14	80

[a] Values obtained by extrapolating information from Reference 4.

Table 4
TYPICAL ENERGY REQUIREMENTS IN TROMMEL SCREENING

Type of Trommel	Nominal throughput (Mg/hr)	Motor size (kW)	Gross specific energy (kWh/Mg)	Separation efficiency (%)
Pretrommel	55	60	1.1	80
Light fraction	1.8	1.5	0.8	80

4. Longitudinal Velocity

The movement of material over or through a screening surface can be characterized by an average velocity, termed the "longitudinal velocity". The adjective "longitudinal" implies motion along the longitudinal axis of the screen. On a flat bed screen, the longitudinal velocity is directed parallel to the screen, and in a trommel screen, along the cylindrical axis.

5. Screen Capacity

Screen capacity is defined as the flow rate per unit area of total screen surface, and is generally expressed as Mg/hr-m^2. Each screening device has a characteristic capacity at a given efficiency or residence time. Screen capacities as related to certain reported efficiencies have been calculated with the use of information gained in research on trommels.[1,4] The results of the calculations are listed in Table 3.

B. Energy Requirements

Calculations indicate that the gross specific energy consumption (i.e., freewheeling plus net work) in pretrommeling with typical equipment should be about 1.1 kWh/Mg, and in trommeling the light fraction, 0.8 kWh/Mg. The throughputs and motor sizes on which the calculations were based are listed in Table 4. When compared to the 5 to 15 kW/Mg gross energy consumption characteristic of shredding and the 3 to 5 kW/Mg in air classification, the energy consumption in trommeling appears quite modest.

C. Design and Screen Performance Parameters

1. Screen Loading

For each particular material (raw refuse or fractions thereof) and set of trommel operating parameters (screen geometry, angular velocity, and inclination), a functional relationship may exist between the screen loading parameter (β, Mg/hr-m^2) and the screening efficiency as illustrated by the slope of the two curves in Figure 9.

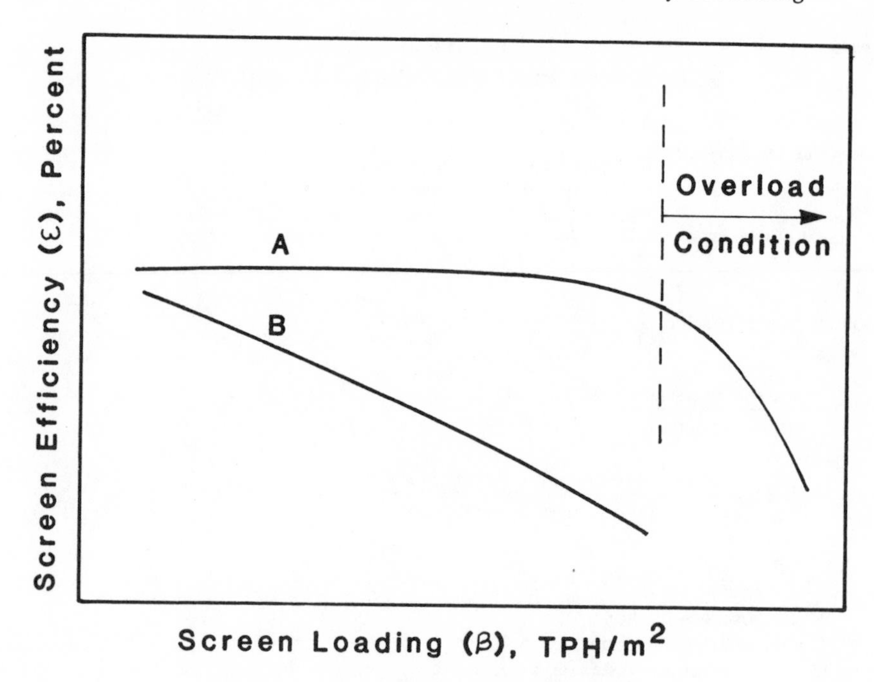

FIGURE 9. Screen efficiency in relation to loading.

The relation indicated by curve A is based on observations made in a study on the trommel screening of air-classified light fraction.[4] The well-defined "knee" in curve A denotes the onset of the overload condition. As the abrupt downward slope shows, overloading is marked by a rapid decrease in screening efficiency. The shape of the loading curve formed by plotting screening efficiency as a function of screen loading, is an important element in the design of trommels. If the curve is relatively level and has a well-defined knee, the application of experimental data to the design of a full-size unit is a fairly straight forward procedure. Mathematically, the scale-up would be determined as follows:

$$S = \dot{m}_d/\beta_o \tag{9}$$

where S is the required surface area of screen (m^2); $\dot{m}_d$, the design throughput, (Mg/hr); and β_o, the value of the screen loading at the point of overload (Mg/hr-m^2). If the design throughput is 40 Mg/hr and the value of β_o is 1.0 Mg/hr-m^2, then the required screen surface area would be 40 m^2.

Through the use of geometrical and centripetal similitude between the experimental unit and the full-size unit along with Equation 9, a complete specification of the geometry and operating conditions of the full-size unit can be made.

If suitable experiments were to be conducted, their results probably would show that the exact form of the loading curve is dependent upon material properties such as composition, particle size, moisture content, and density. Curve B in Figure 9 represents one such possibility.

2. L/D Ratio

The length-diameter ratio (L/D) may be used to characterize the geometry of a trommel screen. The ratio is a convenient parameter because it is dimensionless and consequently is readily applied in scale-up situations. Typical L/Ds range from 2:1 to 5:1

Table 5
SCREENING PERFORMANCE PARAMETERS

Heating value recovery efficiency: $\dfrac{HV(+)}{HV(-) + HV(+)}$

Ash removal efficiency: $\dfrac{ash(-)}{ash(+) + ash(-)}$

Component recovery factor: $\dfrac{C_i(+)}{C_i(+) + C_i(-)}$ or $\dfrac{C_j(-)}{C_j(+) + C_j(-)}$

Mass split: $\dfrac{\text{mass of oversize}}{\text{mass of undersize}}$

Screening efficiency: $\dfrac{\text{material recovered in undersize fraction}}{\text{total undersize in feed material}}$

Note: HV is the heating value;
+ denotes oversize fraction;
− denotes undersize fraction;
C_i and C_j are specified waste components (e.g., paper, plastic, ferrous scrap, etc.).

for processing raw municipal solid waste and the screened air-classified light fraction. Information on the effect of the L/D on trommel operation and performance and on the optimization of the L/D is virtually nonexistent.

3. Angular Velocity

The angular velocity of a rotating trommel screen determines the particle motion and concomitant travel velocity of material along the axis of the screen at a given screen geometry and inclination. The angular velocity must be less than the critical value (assuming only limited slippage of material at the screen surface) for the tumbling and axial movement of the material through the trommel. Under such a condition, (i.e., angular velocity greater than the critical value), the radial acceleration is surpassed by the gravitational acceleration. For refuse processing, typical angular velocities range from 20 to 80% of the critical value with the predominate values occurring between 30 and 50% of the critical value. As of this writing, data describing the effect of angular velocity upon the performance of trommel screens in processing raw refuse or processed fractions thereof were not to be found in the literature.

D. Screen Performance

Items that can be used advantageously in evaluating trommel screen performance are listed in Table 5.

IV. MATERIAL CHARACTERISTICS

Material characteristics of importance in trommel screening are component size distribution, bulk density, moisture content, composition, composite size distribution, and "shredibility" of the material.

The size distribution of the various components of municipal solid waste can be established through screening and hand sorting the material. A plot of cumulative percent passing as a function of screen size represents the size distribution of a particular component. Knowing the size distribution of the individual components, it is possible to select a screen with openings of a size such that a desired component(s) may be retained or passed through the screen, as the case may dictate, and thereby procure a separation of that component(s) from the waste stream.

Bulk density is determined by weighing a known volume of the refuse. The significance of bulk density stems from the fact that bed depth is inversely proportional to bulk density under a given set of operating conditions for the screen.

The importance of moisture content of refuse is in its influence on bulk density, and in its indirect effect on screening efficiency. The latter arises from the tendency of moist, fine particles to adhere to the larger particles of paper and plastics.

The composition of refuse is important because it represents the diversity of the mixture from which a selected component is to be separated, as well as, of course, the abundance of that component. Composition affects the mass split of over- and undersize materials, together with the quality and yield of the screened material. Composition, when interpreted in terms of nature of the material, has an importance in that it then determines the facility with which a material or materials can be separated one from the other. For example, an aggregate of 7-cm rock and 0.8-cm pebbles can be separated into the two components far more readily than can be shredded paper or plastic from the glass fines that tend to adhere to them.

The size distribution of a heterogeneous mixture of refuse is determined through mechanical screening in the laboratory in the manner described in the section on component size distributions. The composite size distribution is a bulk characteristic of the mixture.

Although not directly linked with fuel beneficiation trommel screening can play an important part in the requirements and, hence, design specifications of downstream processing equipment. In particular, the nature of the over- and undersize trommel fractions can have an important bearing on energy consumption and hammer wear in downstream shredding operations. Consequently the trommeled material should be characterized according to the ease with which it can be shredded (i.e., its "shredibility"). For example, pretrommel oversize material consists mostly of cellulosic matter, and as such tends to be highly abrasive and involves a relatively heavy expenditure of specific energy in its size reduction.[6]

The energy expended on shredding the screened light fraction, the air-classified light fraction, and raw municipal solid waste is indicated by the set of three curves in Figure 10. The curve for the screened light fraction closely approximates that to be expected for the pretrommeled fraction. The heavy expenditure of energy involved in the shredding of trommel oversize material is manifested by the relation of the curve for the screened light fraction to those for the air-classified light fraction and the raw waste.

The modeling of pretrommel and posttrommel processes would not be complete without an assessment of the "shredibility" of the oversize material. The reason is that it would be impossible to develop an energy balance and economic data for a trommel system without giving due consideration to material characteristics that may affect the downstream shredding processes. By relying upon data developed by experimentation[7] and having recourse to data on the composition of the trommel oversize material, it becomes possible to make an assessment of the "shredibility" of the various trommel over- and undersize fractions.

V. LIFTERS

"Lifters" is the name given to the protuberances to be found on the inner surface of a trommel screen. They facilitate the movement of material through the screen and aid in

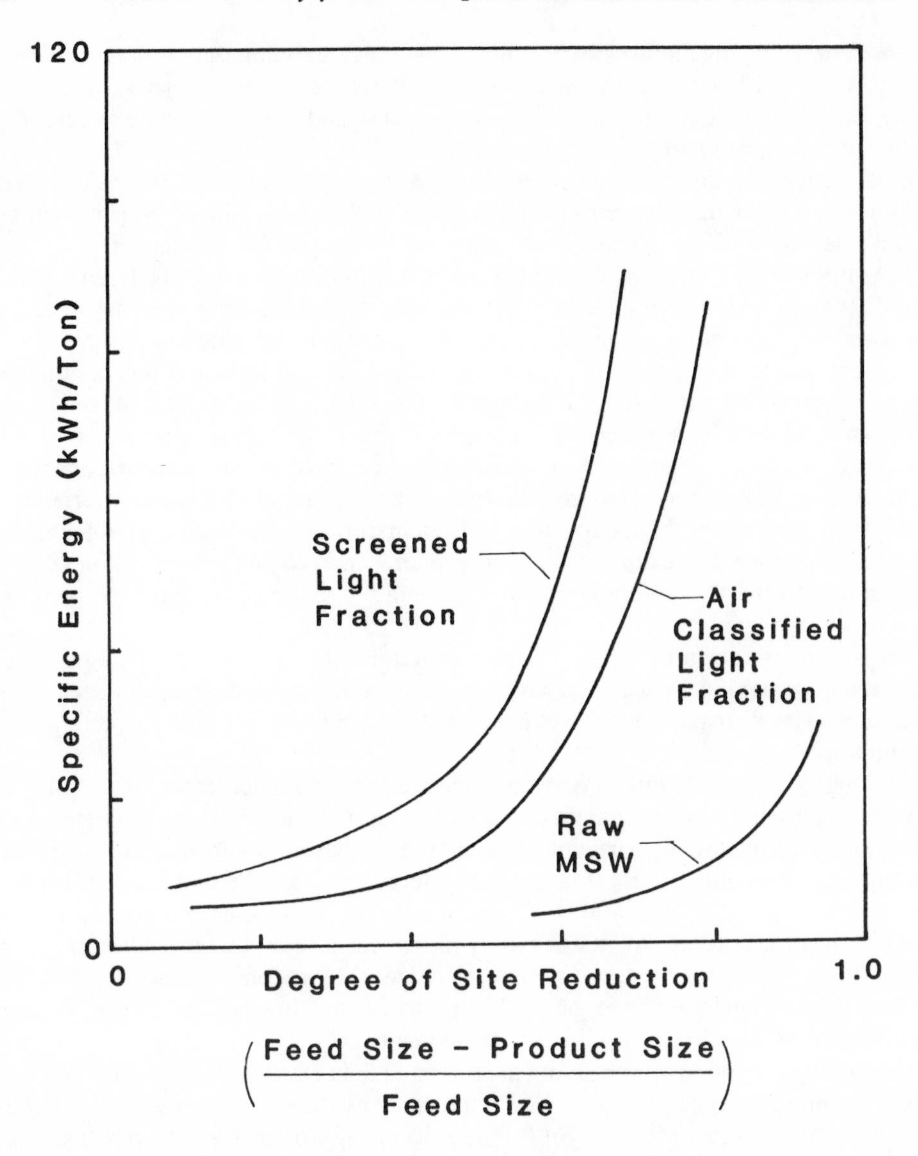

FIGURE 10. Energy consumption in the shredding of raw and of processed MSW.

breakup of refuse bags and containers. Breaking, i.e., rupturing, the bags and containers frees entrapped materials for eventual separation further down the screen. It should also be noted that the act of breaking up the bags and containers implies that a certain degree of size reduction occurs in pretrommeling. Lifters typically are raised spikes arranged in various patterns within the screen or angle or channel iron mounted full-length along the longitudinal axis of the screen. The function of lifters in a trommel screen in which size-reduced refuse is processed is mainly to assist in moving the material through the screen to increase throughput capacity and to accentuate the tumbling motion of the material and thereby enhance screening efficiency. In pretrommels lifters can also serve the important service of lifting glass containers high enough within the screen to promote breakage upon descent and impact with the interior surfaces of the screen. In this manner the glass particles report almost exclusively to the undersize fraction, provided the size of the screen openings has been chosen correctly.

VI. SCALE-UP CONSIDERATIONS

Scaling factors can be derived by way of geometrical and centripetal similitude. Geometrical similitude is established from the optimum length/diameter ratio (L/D ratio) and the required screen area, S. Geometrical similitude is applied for the purpose of determining screen diameter and length, and is done as follows:

$$L/D = K \text{ (value of K established with an experimental unit)} \quad (10)$$

$$S = \dot{m}_d/\beta_o \text{ (value of } \beta_o \text{ determined from experiments with the unit)} \quad (11)$$

$$\pi DL = S \text{ (equation for surface area of cylinder)} \quad (12)$$

$$D = S/\pi L \text{ (rearranging Equation 12)} \quad (13)$$

$$= (\dot{m}_d/\beta_o)/\pi DK \text{ (substituting for S and D)} \quad (14)$$

$$D^2 = \dot{m}_d/(\pi\beta_o K) \quad (15)$$

$$D = (\dot{m}_d/(\pi\beta_o K))^{1/2} \quad (16)$$

$$L = K\,(\dot{m}_d/(\pi\beta_o K))^{1/2} \quad (17)$$

Centripetal similitude follows from the required force balance to maintain similarity of particle motion between the experimental and the full-size trommel. The similarity of forces implies that the rim velocities (radius times the angular velocity) of the full-size unit ($\omega_2 D_2/^2$) equal that of the experimental unit ($\omega_1 D_1/^2$), that is,

$$\omega_2 D_2/2 = \omega_1 D_1/2 \quad (18)$$

$$\omega_2 = \omega_1 D_1/D_2 \quad (19)$$

From Equations 16 and 19 are derived the required angular velocity of the full-scale trommel (ω_2):

$$\omega_2 = \omega_1 D_1/(\dot{m}_d/\pi\beta_o K)^{1/2} \quad (20)$$

$$= \omega_1 D_1(\pi\beta_o K)^{1/2}/\dot{m}_d^{\ 1/2} \quad (21)$$

Equations 16, 17, and 21 express the required geometry and angular velocity of the scaled-up trommel in terms of those quantities for the experimental unit. Lastly, the angle of inclination (α) for the full-scale trommel will be identical with that of the experimental unit.

Inasmuch as the scale-up factors are a function of properties of the throughput material (e.g., particle size, composition, density), they most likely will be different for each of the various refuse fractions (e.g., raw, shredded, air classified).

REFERENCES

1. **Bernheisel, J. R., Bagalman, P. M., and Parker, W. S.,** Trommel procession of municipal solid waste prior to shredding, paper in Proc. 6th Miner. Waste Util. Symp., U.S. Bureau of Mines and IIT Research Institute, Chicago, May 2 and 3, 1978.
2. **Savage, G., Diaz, L. F., and Trezek, G. J., RDF:** quality must precede quantity, *Waste Age*, 9(4), 100, 1978.
3. **Trezek, G. J. and Savage, G.,** MSW component size distribution obtained from Cal resource recovery system, *Resour. Recovery Conserv.*, 2, 67, 1976.
4. **Savage, G. and Trezek, G. J.,** Screening Shredded Municipal Solid Waste, *Compost Sci.*, 17(1), 7, 1976.
5. **Woodruff, K. L.,** Preprocessing of municipal solid waste for resource recovery with a trommel, *Trans. Soc. Min. Eng.*, 260, 201, 1976.
6. **Hill, R. M.,** Rotary screens for solid waste, *Waste Age*, 18, 33, 1977.
7. **Savage, G. and Trezek, G. J.,** Significance of Size Reduction in Solid Waste Management, Vol. 2, report to the U.S. EPA under Contract No. R-805414-010, Municipal Environmental Research Laboratory, Cincinnati, Ohio, EPA-600/2-80-115, August, 1980.

Chapter 7

MATERIAL RECOVERY

I. INTRODUCTION

Although in the past 5 years the reclamation of useful materials from urban wastes may have fallen behind energy recovery in the public attention, much can be said in favor of restoring it to its original primal status or at least to one equal of that of energy recovery. Certainly, materials recovery can constitute the deciding factor in the economic feasibility of an energy recovery undertaking that by itself would at best be only marginally feasible. Furthermore, the arguments advanced in the past for the conservation of our vanishing natural resources through the reclamation of useful materials continue to hold true, despite the current federal policy of downplaying those arguments.

This chapter deals solely with paper, glass, aluminum, and magnetic metals, even though other useful materials may be found in municipal solid wastes (MSW). The reasons for the restriction are fourfold: (1) the four materials are present in MSW in amounts sufficiently large to warrant the investment of the money and effort needed to recover them; (2) while not always constant, a market usually can be found for them; (3) the technology for their recovery is reasonably well developed, although that for aluminum recovery leaves much to be desired; and (4) the fourth reason is specific to this book: space limitations restrict a detailed presentation to only the most important of the materials to be found in the wastes, i.e., to the four discussed in this chapter.

II. PAPER

A. The Need to Recycle Paper Fiber

As was stated earlier, paper and paper products make up the largest fraction of the municipal solid wastes generated in the U.S., constituting as they do from 40 to 50% of the daily output of wastes. The existence of the range of values is a function of differences between locales in terms of economic, social, and nature of the pursuits of the dwellers in the locales. Whether or not paper and paper products will continue to be the major component of urban wastes into the indefinite future remains to be seen. Nevertheless, it is safe to conclude that an essential factor in the continued high rate of consumption of paper and paper products is a rigorously prosecuted program of reclamation and recycling of the paper discarded into the waste stream. The lack of concern about this grim necessity is shown by the fact that as late as 1977 and 1978 only about 25% of the paper consumption was being recovered.[1,2] Interestingly, the higher rates of reuse (29.3 and 25.1%) were in the northeastern and north central states, while the lower rates (15.3 and 16.5%) were in the southern and western states.[2]

Despite the fact that the cellulose in the form of cellulosic fibers that make up paper is a renewable resource, there is a limit on the rate at which it can be renewed. The limitation is in greater part determined by that of the availability of appropriate land and energy for nonnutritional crop production. Moreover, the strong possibility is that another competing factor will come into play in the not too distant future. It is the potential aggravation of the limitation on land availability brought about by the proposal of yet another competing use of arable land, namely, the production of biomass for energy production.

B. Uses for Reclaimed Paper

Reclaimed paper fiber can be used: (1) as a fiber in the manufacture of paper, paperboard, or other paper products; (2) as a fuel in energy production; (3) as a feedstock in various fermentations; and (4) as a bulking agent and carbon source in compost production. An important feature that distinguishes the utilization of reclaimed paper in the production of paper and paper products from the other uses is that it depends mainly on paper that has not been contaminated through entry into the solid waste stream. Inasmuch as the use of paper as a fuel, as a feedstock in fermentation, and as a carbon source are covered in other sections, the only type of utilization discussed in this section is its role as a fiber in the manufacture of paper and paper products.

A case based on a lower energy requirement could be made for the use of virgin fiber under certain circumstances in paper manufacture. Thus, it has been pointed out that integrated mills built and operated to produce paper from trees can obtain a sizeable portion of their energy through the burning of by-products. As a result, it often happens that such mills need to bring in less energy than do those that manufacture paper from wastepaper.[3] The latter is true because recycling mills often are located in relatively densely populated areas, and hence must obtain their energy from relatively clean fuels, e.g., gas, oil, or coal. However, a portion of the energy requirements of recycling mills could be met through the use of solid waste as a supplementary boiler fuel.[4]

C. The Wastepaper Industry

The principal user of reclaimed paper fiber is the wastepaper industry. Therefore, at this time a brief survey is made of the wastepaper industry as a whole so as to more properly establish the potential of paper recovery from mixed municipal solid wastes. The paper stock industry is the principal vehicle for the recycling of paper. The nature of the industry is indicated by the interrelationship between types of wastepaper producers and users as charted in Figure 1. The wastes are funneled through the collecting, sorting, grading, packing, and shipping processes which come within the overall organization of the paper stock industry.

It should be noted that the chart does not refer to the paper in mixed wastes, i.e., the municipal solid waste stream, nor are they included in the two subsequent sections on sources and specifications. The treatment on mixed wastes is reserved for a later section.

D. Sources of Wastepaper

Examples of types of wastepaper from the converting plants charted in Figure 1 are envelope cuttings, manifold form cuttings, tabulating card cuttings, paper bag cuttings, and boxboard and corrugated cuttings. The waste from newspaper companies is in the form of white news blanks, mill wrappers, and overissue newsprint. Used corrugated containers make up the principal paper waste from industrial plants. Supermarkets and other stores discharge a large number of used corrugated containers and a lesser amount of mixed paper. The output from office buildings is primarily tabulating cards, computer printout, forms, and an assortment of paper. The major output of paper from residences is in the form of newspapers. (The remaining output of paper from residences, i.e., assorted circulars, magazines, and junk mail, is discarded with other household wastes as refuse.)

In terms of quantity, the largest share of reclaimed fiber comes from corrugated products. About a third of the corrugated paper comes as cuttings from convertor (i.e., box) plants, and the remaining two thirds is in the form of used boxes from supermarkets, retail stores, and industrial plants.

The second major source of reclaimed fiber is the high-grade scrap and de-inking

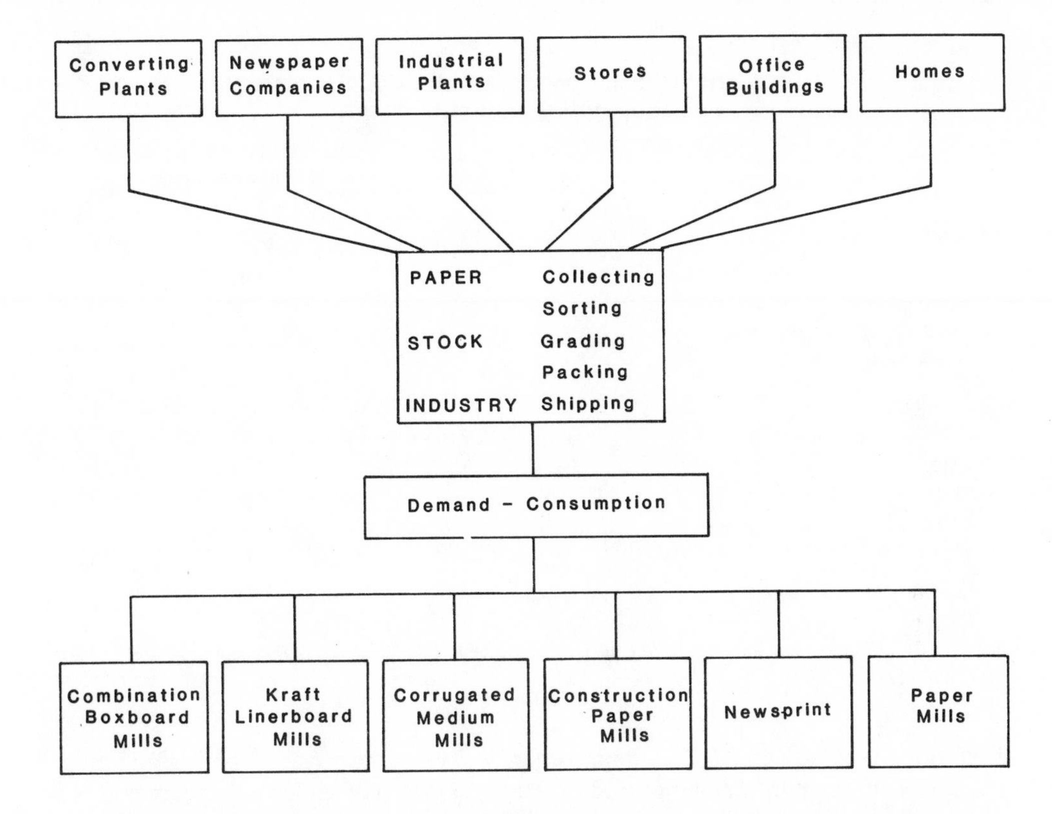

FIGURE 1. Wastepaper generators and potential market areas for reclaimed paper fiber.

paper collectively termed "pulp substitutes". Most of the high-grade reclaimed paper originates in printing and converting plants within the industry. Generally, the cuttings from such plants are sorted, baled, and shipped to recycling mills. Mixed grades and newspaper, respectively, constitute the third and principal sources of used fiber. Mixed grade paper is mostly office waste, although boxboard cuttings from folding carton plants constitute about 30% of the mixed paper output. The major share of reclaimed newsprint fiber comes from newspapers collected by community organizations, although a few municipalities have established newspaper collection.

Types of wastepaper used as feedstock in combination boxboard mills are bleached and unbleached high grade paper stock, deinking grades, corrugated Kraft, overissue and No. 1 newsprint, mill wrappers, boxboard cuts, and mixed paper. Kraft corrugated and corrugated containers are used in Kraft linerboard and in corrugated medium mills. White newsblank, corrugated containers, No. 1 newsprint, and mixed paper serve as raw materal in construction paper mills. Especially sorted newsprint is deinked and processed for reuse in mills designed for that purpose. Lastly, high-grade paper stock is manufactured into paper in conventional papermills.

Regardless of source, with each pass from 20 to 25% of the paper fibers are rendered unsuitable for reuse in paper manufacturing because the fibers become too short. However, it has been reported that after an initial 20% loss of strength, newsprint apparently can be recycled repeatedly.[5]

1. Specifications

To ensure a modicum of order in marketing, the Paper Stock Institute of America publishes a set of specifications for 46 identified grades of wastepaper. Depending upon their mix of business, the various dealers handle different combinations of grades. Some

Table 1
GRADES OF WASTEPAPER AND THEIR 1976 MARKET VALUES

Item	1976 Market value ($/Mg)
Tab cards	
Boxed	127
Loose	91
Computer print-out	86
White ledger (white office type wastes, having a limited amount of ink and color printing)	45
Colored ledger	27
Corrugated	23
Newspaper	18—23
Mixed paper waste, baled[a]	5—9[b]

[a] Loose mixed paper must be baled at the local dealer's plant before the material can be moved. It is accepted only when market conditions are very favorable.
[b] When the market is strong, it may go up to $18/Mg.

Modified from *Paper Stock Standards and Practices, Circular PS-74,* Paper Stock Institute of America, New York, 1974.

of the more common grades include mixed paper, news, corrugated, high-grade pulp substitutes, high-grade deinking grades, etc., examples of which and their market values are listed in Table 1.

A complete list of specifications is given in the publication *Paper Stocks, Standards and Practices, Circular PS-74*.[6] The following specifications of selected pertinent grades are given to serve as a point of reference:

1. No. 2—mixed paper
   A. Consists of a mixture of various qualities of paper not limited as to type of packing or fiber content.
      1. Prohibitive material may not exceed 2%
      2. Total-out throws may not exceed 10%

2. No. 1—mixed paper
   A. Consists of a baled mixture of various quantities of paper containing less than 25% of ground wood stock coated or uncoated
      1. Prohibitive materials may not exceed 1%
      2. Total-out throws may not exceed 5%

3. Super mixed paper
   A. Consists of a baled clear sorted mixture of various qualities of papers containing less than 10% of ground wood stock coated or uncoated
      1. Prohibitive materials may not exceed ½ of 1%
      2. Total-out throws may not exceed 3%

4. News
   A. Consists of baled newspaper containing less than 5% of other papers
      1. Prohibitive materials may not exceed ½ of 1%
      2. Total-out throws may not exceed 2%

5. Super news
   A. Consists of baled sorted fresh newspapers, not sunburned, free from papers other than news, containing not more than the normal percentage of rotogravure and colored sections
      1. Prohibitive materials—none permitted
      2. Total-out throws may not exceed 2%

6. Corrugated containers
   A. Consists of baled corrugated containers having liners of either jute or Kraft
      1. Prohibitive materials may not exceed 1%
      2. Total-out throws may not exceed 5%

7. No. 1 sorted colored ledgers
   A. Consists of printed or unprinted sheets, shavings and cuttings of colored or white sulfite or sulfate ledger, bond, writing, and other types of paper which have a similar fiber and filler content. This grade must be free of treated, coated, padded, or heavily printed stock
      1. Prohibitive materials—none permitted
      2. Total-out throws may not exceed 2%

8. No. 1 sorted white ledger
   A. Consists of printed or unprinted sheets, shavings, and cuttings of white sulfite or sulfate ledger, bond, writing, and other papers which have a similar of treated, coated, padded, or heavily printed stock
      1. Prohibitive materials—none permitted
      2. Total-out throws may not exceed 2%

E. Demand (Market) for Wastepaper

During the past two decades, the extent of paper recovery from the waste stream has fluctuated only a few percent from year to year since 1960, i.e., from a low of 21.2% in 1965 to highs of 24.7 and 24.4% in 1977 and 1978. These percentages include paper collected for export and for other uses.[1,7] In Figure 2 are plotted the price indexes from 1965 to 1974.[8] The curves in the figure show that the fluctuations in the price index for used paper have not reflected the relative stability of percentage recovery. The sharp fluctuations which occurred in the 1972 to 1975 period persisted through the remainder of the decade. However, the potential competition for the use of fibers and other biomass for other uses (e.g., energy, chemical synthesis) has led to the optimistic consensus that the fluctuations will be replaced by a continued strong demand through the 1980s.

F. Recovery from Mixed Solid Wastes

Aside from the newsprint and corrugated removed prior to reaching the disposal or processing site, the specifications for mixed paper grades are the most applicable to the air-classified, screened light fraction recovered from the waste stream. If the light fraction is further processed through an especially designed fiber recovery process, the fibers can be upgraded to a level at which they meet the specifications of a pulp substitute. Examples of upgrading processes are the Cal Recovery Systems (CRS) fiber recovery process,[9,10] one proposed by Yoda et al.,[11] another by Nollet and Sherwin,[7] and yet another by the National Center for Resource Recovery.[12,13] The Black Clawson process often is mentioned as an example of a wet system for upgrading fiber to paper

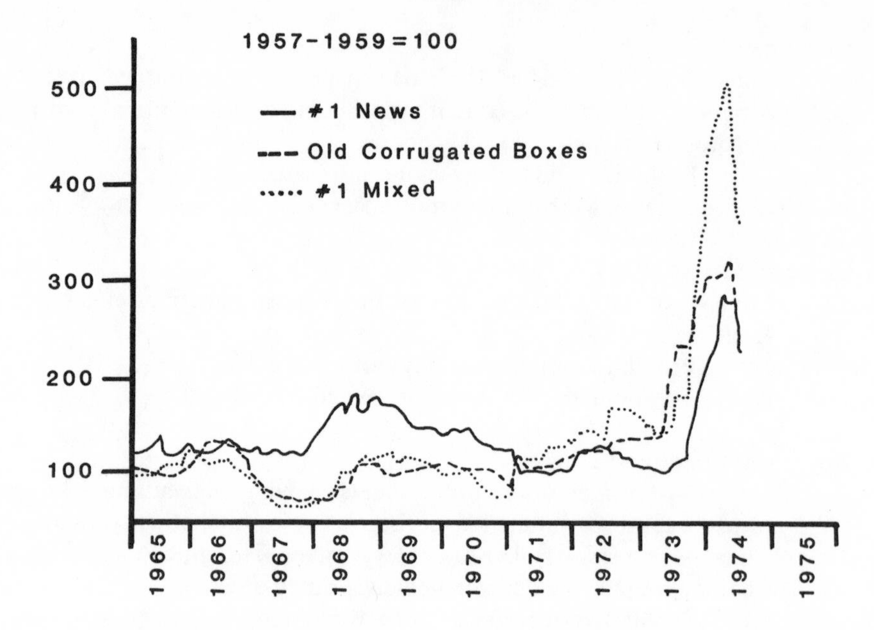

FIGURE 2. Price index for the three major wastepaper grades—1965 to 1974.

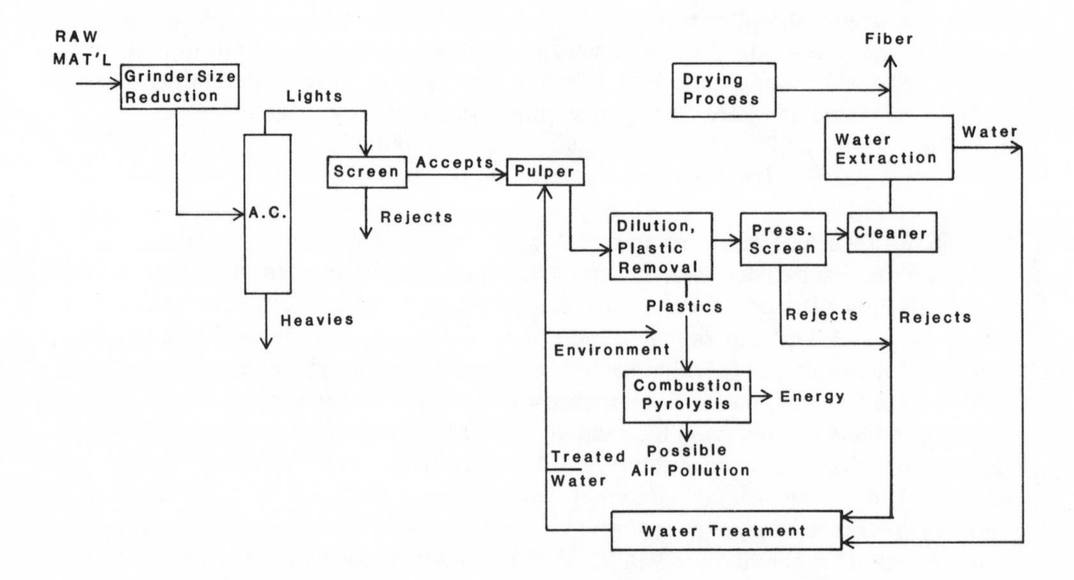

FIGURE 3. Cal Recovery Systems (CRS) paper fiber recovery process.

manufacture quality. However, up to and including the time of this writing, fiber produced with the use of the Black Clawson system has been of an inferior quality.

1. CRS Process

The Cal Recovery System (CRS) process is selected for description because it involves the use of "off-the-shelf" equipment, combines dry and wet processing, and has been attended by successful results. A flow diagram of the system is given in Figure 3. As the figure indicates, the system begins with a front-end dry processing stage in which the incoming wastes are size reduced and air classified. The lights are passed through a trommel screen, the "accepts" of which contain the paper fiber to be reclaimed.

The wet processing stage begins with the pulping of the screened lights to form an approximately 3% slurry. At a solids content approximating 3%, the fibers become slurried, whereas plastic contaminants retain their relatively large particle size. A conventional ragger can be incorporated into the pulping operation. The removal of plastics is accomplished directly after the pulping step by discharging the pulper contents to a screen designed to remove the plastic particles. The fiber slurry is further diluted to form a 0.5% solids slurry. The dilution is required to render the slurry amenable to centrifugal cleaning. Centrifugal cleaning is accomplished in two steps. In the first step, very fine glass, grit, and dirt contaminants together with fiber bundles and some usable fibers are removed. These materials exit through the bottom of the cyclone with which the centrifugal cleaning is being accomplished, while hot melt, wax, and some remaining fine plastics are discharged through the center probe of the cleaner. Since it contains usable fibers, a separate step is used to clean the reject stream. The accepts from the reject cleaner are combined with the first stage cleaner accepts. The rejects from the reject cleaner are disharged from the system and constitute a fiber loss of about 5% of the total fiber entering the system. A similar procedure is followed in the second cleaning step. However, because the center probe and bottom rejects are considerably less contaminated than those from the first step, the stream is returned to the beginning of the first stage and reprocessed. In some cases, it may be desirable to include a large holding tank or pond between the first and second cleaning steps. This would allow dirt loosened in the centrifugal cleaner and carried with the accepts to be floated off and thereby removed from the system. A noticeable brightening of the fiber takes place after the first stage of the cleaning process.

After the two cleaning steps, only an occasional small contaminant is present. The fiber produced in the two steps is suitable as a secondary stream for media and liner board manufacture. The addition of yet another cleaning step would lead to the production of a fiber slurry having only occasional speck contaminants. The final step in the fiber processing involves the use of conventional dewatering, and possibly of drying and baling for shipment. All of the process water is recirculated within the system. After each pass, 25% of the water is subjected to a conventional primary and secondary activated sludge waste-water treatment process. Make-up water is introduced at the pulping stage. An idea of the quality of the resulting pulp may be gained by comparing its sheet properties with those of pulps derived from unmixed wastepaper sources as listed in Table 2.

G. Public Health Constraint

A major constraint on the use of fibers reclaimed from mixed waste is constituted by Food and Drug Administration's (FDA) restrictions in force at the time of this writing. The FDA Food Additive Regulation 21 CFR S176.260, "Pulp from Reclaimed Fiber", prohibits the use of paper for food, drug, or cosmetic containers if such a paper had been used for shipping or handling poisonous or deleterious substances. The use of paper containing more than 10 ppm of PCB is forbidden in another FDA regulation. The practical consequence of regulation 21 CFR S176.260 is the preclusion of the use of all paper fibers extracted from the mixed waste stream for the manufacture of containers for food or cosmetics. The reason is readily apparent, namely, it would be impractical, if not impossible, to determine the prior uses of paper extracted from the mixed waste stream.

In view of the nature of mixed municipal solid waste, the existence of a concern about the presence of disease-causing microorganisms in the paper made from reclaimed

Table 2
PROPERTIES OF PULPS FROM PROCESSED MIXED WASTES AND FROM UNMIXED WASTEPAPER STOCKS

Source[a]	Burst factor (m^2/cm^2)	Tear factor (m^2)	Breaking length (m)
100% De-inked newspapers[b]	9.3	38.3	2861
Spruce groundwood[b]	10.7	44.2	2610
Mixed waste—residential	15.6	132.0	2325
Mixed waste—commercial	17.9	138.0	2765
Newsprint furnish[b]	15.4	82.4	3300

[a] Based on tests involving "handsheets" made according to *Tappi* standards.
[b] According to Klungness, J. H., *Tappi*, 58(10), 128, 1975.

fibers is not surprising. As was pointed out in another chapter, the concentrations of enteric indicator organisms in raw municipal solid wastes rivals those in raw sewage sludge. At the same time, it should be emphasized that in all probability practically all of the organisms are of nonhuman origin. Nevertheless, appreciable concentrations in the reclaimed product are not to be ignored.

Work done on the microorganism content of paper and paper products made from fiber reclaimed from mixed wastes has been very limited in extent. Renard[13] reports a total absence of fecal *Streptococcus* and coliforms from paper rolls from mixed waste feedstocks, although the count of total bacteria ranged from 482 to 2356 organisms per square centimeter of paper. In Japan, Yoda et al.[11] brought the coliform count down from 10^6 to 10^8/g in the raw material to 10^3/g in the reclaimed pulp by adding hypochlorite to the pulp. The addition of the hypochlorite had the added advantage of improving the color of the pulp. They also analyzed the reclaimed pulp for an assortment of toxic organics and heavy metals. While their paper fails to give concentrations, they do state that the Cd and As contents of the pulp were lower than those in the municipal refuse, but higher than in ordinary wastepaper.

Renard[13] summarizes the microbiological situation quite well. He concludes that although bacteriological contamination could be a potential technological barrier in the industrial paper mill environment, the total absence of enteric indicators in the product are a favorable indication of the lack of contamination hazards in the use of fiber reclaimed from mixed waste.

III. GLASS

Glass constitutes about 10 to 11% of the municipal solid waste (MSW) stream. Of this fraction, about 90% is in the form of flint (clear), amber, and green bottles and jars.[14] Glassware and plate glass constitute the greater part of the remainder.

A. Rationale for Removal of Glass

Reasons for removing glass from MSW in a resource recovery operation are twofold. The first and at present perhaps the overriding reason is to improve the quality of some other component that is being removed or processed. The second, lesser reason is to reclaim the glass for reuse. Two commonplace examples of the first reason are enhancement of the heating content and general fuel characteristics of RDF, and improvement of the quality of a compost product. Glass not only lowers the heating value of RDF, it also leads to the development of serious problems in boiler

Table 3
AMOUNT OF RECLAIMED GLASS PERMITTED IN MANUFACTURE OF CERTAIN ITEMS

Item	Reclaimed glass[a] (%)	Specifications
Brick	50	No metals or organics
Concrete aggregate	50	No metals or organics
Foamed glass	95	Needs no cleaning, sorting
Ceramic tile	40—60	Sized to −5, +200 mesh
Terrazzo tile	60—70	No metals
Building panels	94	No metals, sized to 200 mesh
Glass wool	10—50	Up to 20% foreign material
Slurry seal	100	Sized from −0.9 cm to 200 mesh, small amounts of metals and organics
Glasphalt	77	Size 0 to −1.25 cm

[a] Fraction of the raw material that may consist of reclaimed glass.

Modified from *Waste Age,* 11(5), 10, 1980.

management. Glass in a compost product lowers the aesthetic quality of the of compost, and may also act as a constraint on the utilization of the product. Glass shards in a mass of compost are conspicuous to a degree out of proportion to their abundance. Convinced that he or she may be injured, the prospective user shies away from any compost that contains glass. It should be pointed out, however, that despite the visibility, chances of being cut or scratched in the course of handling glass contaminated compost generally are quite remote, principally because sharp edges are abraded or polished during the course of the composting process. On the other hand, the glass would pose a danger to grazing animals if the compost were spread on pastures.

B. Uses for Recovered Glass

Proposed and actual uses for reclaimed glass cover a wide range of utility and potential for monetary return. They range from serving as an aggregate in asphalt pavement to cullet in the manufacture of glass containers and flat glass. An idea of the potential demand for the glass can be gained from the data in Table 3. As an aggregate, the value of recovered glass is only from \$2 to \$5/Mg. At the other extreme, its value could be on the order of \$30/Mg when used as a cullet. Between the two extremes is use in brick manufacture and in the manufacture of fiberglass. For brick manufacture, value of reclaimed glass would be on the order of \$8/Mg,[15] and in fiberglass manufacture, about \$18/Mg.[16]

For each use, the recovered glass must meet specifications dictated by the nature of the use and by the user. Because of this fact, the range of specifications matches that of the potential uses. The extent to which a given glass meets these specifications is determined by testing according to standard methods established by the American Society for Testing and Materials (ASTM).[17]

1. Aggregate

Among the applications in which suitably prepared glass can serve as an aggregate, one that has received a considerable, though passing notice, is use as an aggregate with asphalt in pavement construction. The mixture of asphalt and glass was given the descriptive name, "glasphalt". Indeed, in the late 1960s and early 1970s, its proponents saw the potential for the use of glasphalt in street and highway paving as being almost limitless.[18-20]

In conventional road construction, aggregate is mixed with asphalt to increase stability of the asphalt because by itself, asphalt has a stability too low for most pavement purposes. The increase in stability depends in large part upon the adhesion of the asphalt to the individual aggregate particles. The adhesion is promoted by the presence of minute pores or depressions in the aggregate particles. Asphalt fills the pores and depressions, thereby forming a bond between the mass of the asphalt and the surfaces of the particles.

For glass to be used as an aggregate in asphaltic mixtures, generally the preference is for nearly equidimensional particles rather than for flat or elongated particles. While this preference can be met, other properties of glass particles are poorly suited for use as an aggregate. For example, waste glass particles are almost nonporous, smooth surfaced, and angular, with a preponderance of flat particles in sizes retained on a No. 4 sieve. The combination of low porosity and smooth texture results in less internal friction in the asphaltic mixture and hence a lower strength or stability. Flat particles would be expected, since the average bottle wall thickness is approximately 2.8 mm, and particles with any dimension greater than three times this value would still be classified as flat and elongated. Any further thinning of the walls of the containers detracts from the usefulness of the particle, since the percentage of flat, elongated particles in the mass of particles would be enlarged. However, compensation for the increase in the percentage can be made by altering the gradation of glass used in the mixture. Another undesirable property is the loss of adhesion between asphalt and glass in the presence of water, unless additives are used to improve adhesion. One such additive is hydrated lime applied in an amount equal to 1% by weight of the aggregate.[21]

An advantage from the use of glass-asphalt mixtures is a reduction in cooling rate due to the glass. The slow cooling rate, even in winter, allows more time for compacting the mixture, and thus results in an improvement in the durability of the glasphalt.

Unfortunately, the many advantages expected from the use of the material have not been forthcoming. This failure, coupled with a much less than favorable economic situation, has resulted in an almost complete loss of interest in the product on the part of the pavement industry and governmental transportation departments.[8] The unfavorable economics are attributable to the necessity of using hydrated lime to ensure adhesion between glass and asphalt,[21] and especially to the high cost of transporting the glass to the point of usage.

A second, and perhaps more promising use for glass is as an aggregate in making concrete.[22-24] An especially promising use in concrete manufacture is as a foamed glass aggregate. Because of the fact that foamed glass lightweight aggregate can be prepared in a wide range of densities, it is feasible for a single plant to produce concrete over a wide range of densities. Light and highly insulative concrete blocks can have a density of 320 to 352 kg/m^3. A density of 160 to 224 kg/m^3 would suffice as a low cost bulk insulation material poured in place to fill concrete blocks or to retrofit existing uninsulated buildings. At the time of this writing, not enough information was available to permit firm conclusions regarding economics of the application.

2. Manufacture of Bricks

The principal use of waste glass in brick manufacture is as a fluxing agent.[25-26] The advantage in the use of glass stems mainly from the lowering of the required firing temperature and a shortening of the firing time otherwise required in the manufacture of bricks and other structural clay products made from common clays. The resulting benefits are a reduction in energy requirements, a lowering of manufacturing costs, and, of course, a recycling of glass.[26] According to Tyrell et al.,[25] the substitution of waste glass for one half the clay in red mixtures reduces the heat required to fire the body to maturity by approximately one million kJ/Mg. With the tan bodies, the reduction is about 900,000 kJ/Mg. Conservation in terms of fuel consumption made possible through the use of glass (i.e., 50:50 glass-clay mixture) in red brick manufacture in a plant that produces 36 million face bricks per year would amount to about 1.818 million cubic meters of 1000-kJ natural gas per year. In the manufacture of tan body bricks, the savings would be about 1.632 million cubic meters of 1000-kJ gas per year. Furthermore, the substitution would result in a 30% increase in production of red body brick without additional kiln capacity. In the production of tan body brick, the increase in production would amount to 23%.

Two important specifications for waste glass intended for use in the manufacture of bricks are (1) no organic material should be in the waste glass; and (2) the glass must be ground to −200 mesh. Organic matter interferes with the fluxing action of the glass. The proposed maximum level for organic material is 10%.

3. Glass Polymer Composite (GPC) Manufacture

Recovered crushed waste glass, when mixed with polymer at an approximate ratio of 87% glass to 13% polymer, can be used to cast glass polymer composite (GPC) materials.[24] On a small scale, GPC pipe has been developed for use as a sewer pipe. The pipe has been demonstrated to have superior compression strength and resistance to sewer acids and gases. According to Willey and Bassin,[24] GPC pipe probably would be competitive with concrete and vitrefied clay pipe up to 9.6 cm in diameter.

4. Foamed Glass Insulation Panels and Glass Wool

Pulverized glass containing impurities has been used in the production of foamed glass insulation panels having densities ranging from 224 to 640 kg/m^3. Materials can be made for specific applications by compositing one or more of a variety of fillers and binders with foamed glass. Walls and doors in which foamed glass panels serve as a core material have been shown to have excellent thermal and acoustic insulation. Moreover, the panel is noncombustible. Glass wool is used principally for thermal insulation against heat and cold, and for sound absorption. Its manufacture involves the impingement of a jet of compressed air on a stream of molten glass. Research conducted at Tuscaloosa Metallurgy Research Laboratory of the U.S. Bureau of Mines showed that glass in incinerator residue could be used to make a glass wool that would meet requirements for commercial use.[27] If the information in *Waste Age*[16] is regarded as being indicative, glass separated directly from the MSW water stream could also serve as a raw material for glass wool manufacture.

5. Cullet

The final use mentioned in this section is as a cullet in glass container production. The demand for and market value of cullet is sufficiently high as to render such a use quite practical if the cost of its production could be kept at a reasonable level. As of 1977[14] the market value of good quality cullet was on the order of $33/Mg. The introduction of cullet results in a lowering of the temperature at which a furnace must be operated in glass production, and air emissions are reduced. The lower furnace temperatures lead to a reduction in energy consumption and a prolongation of the refractory linings of the furnace. By way of a comprehensive energy balance, Lubisch[28] shows that an energy savings of 951 kJ/kg can be made by substituting cullet for primary raw material. He estimates that for each 10% increase in cullet, there would be a corresponding 2.2% decrease in fossil fuel required, a fuel savings of 4.4% and an electric energy savings of 1.1%.

The existence of the favorable conditions mentioned in the preceding paragraph would lead one to expect that the greater part of the glass in the municipal solid waste stream would be used as cullet. The fact is that two important factors stand in the way of such a use: (1) cost of transporting the waste glass to the site of use; and (2) lack of an adequate technology for processing the waste glass to meet the rather rigid specifications for cullet.

Glass recovered from MSW has two very objectionable characteristics that must be removed or changed before the material can be used as cullet in glass manufacture. They are the presence of so-called "refractory" particles and the inclusion for chemical compounds that would determine the color of the new glass product. The presence of iron and chrome oxides imparts an amber or green color to glass. Flint (clear) glass contains no coloring chemicals or at the most, in concentrations not great enough to impart a tint. Certain contaminants (e.g., iron oxide particles) also can result in a tinted product. Consequently, in the manufacture of glass from waste glass cullet, certain specifications regarding color have been established for the cullet. Flint cullet should contain no more than 0.06% iron oxide (Fe_2O_3) and 0.001% chrome (Cr_2O_3). Cullet for amber glass should contain no more than 0.025% Cr_2O_3, and for green glass, no more than 0.2% Cr_2O_3. Thus, cullet for use in flint glass manufacture should contain no more than 5% amber glass. Moreover, the magnetic metal content of the cullet must be very low.

The problems of color is further complicated by the existence of a considerable colorant variability with respect to geographical area. The extent of the variation is indicated by the data in Table 4. As is pointed out by Duckett,[15] the little information that is available for geographical color variations is reported in terms of averages for the individual areas and only on the basis of a very limited sampling. What is now needed is information on the variability within the geographical area itself. Thus, variability of colorants in colored mixed MSW cullet must be regarded, along with the mean level of colorant, as a barrier to the widespread use of MSW cullet.

The second major obstacle, namely, lack of a practical technology for the satisfactory processing of MSW cullet, has been the object of a more or less intensive effort during the past decade. Despite the many claims for seeming "breakthroughs", at the time of this writing the problem remains as formidable as ever, as will become amply evident upon reading the section on "Methods" which immediately follows this paragraph. However, it should be emphasized that this rather negative statement applies only to recovery for use as cullet. The existing technology for processing for the less exacting uses (e.g., as an aggregate, in brick manufacture) is mostly satisfactory.

C. Methods for Recovering and Processing MSW Glass

An obvious means of separating glass from the waste stream is source separation. The advantages of source separation, especially with respect to glass recovery are many, as are the disadvantages. A discussion of them and of the ways of initiating and carrying out a program of source separation is beyond the spatial limitations of this book. Nevertheless, it should be noted that at the time of this writing, the recovery of postconsumer container glass is almost entirely accomplished through source separation. Attention also is called to a form of source separation that could, with suitable legal and motivational efforts, be put into practice within a short period of time. It is the institution of statewide program for the use of returnable beverage bottles and the simultaneous abandonment of the use of nonreturnable bottles. The effect of such a move on the rate of generation of glass waste is indicated by the fact that from 80 to 90% of the glass in MSW is in the form of beverage and food containers. A major benefit would be a drop in energy consumption, since the energy involved in processing a

Table 4
GEOGRAPHICAL VARIATION IN COLORANT CONTENT OF MSW CULLET

	Color mix (% wt)			Colorant conc (% wt)	
	Clear	Green	Amber	F_2O_3	Cr_2O_3
Franklin, Ohio[a]	61	2	36	0.228	0.0078
Palo Alto, Calif.[a]	69	22	9	0.137	0.011
San Francisco, Calif.[a]	67	23	10	0.106	0.048
Estimated U.S. average[b]	65	15	20	0.108	0.022

[a] From reference 21.
[b] From reference 15.

returned container for reuse is a small fraction of that required to produce a similar container from raw material.

1. Mechanized Removal of Glass

In a mechanical resource recovery operation, practically all of the glass in the MSW input finds its way into the heavies in air classification. Glass is further concentrated by the succeeding treatment accorded to the heavies. The flow succession of interest in terms of glass recovery begins with the treatment of the "accepts" from the trommel screening to which the heavies are subjected. A flow diagram of a very effective method of concentrating the glass in the trommel rejects and developed by Cal Recovery Systems, Inc. (CRS) is shown in Figure 4. In essence, a "stoner" is an inclined vibrating screen through which air is forced, as a result of which material on the screen is "fluidized". Heavier particles (mostly glass) in the fluidized mass move in one direction, and the lighter, in the opposite direction.

The heavier fraction is then subjected to the screening and fluidizing steps diagrammed in Figure 4. The glass content of the material in the "accepts" of the fluidization is sufficiently great to permit the material to serve as a feedstock in the manufacture of fiberglass and other glass products that require a raw material of a comparable specification. It does not quite meet the usual specification for a cullet for use in container glass manufacture.

2. Processing to Meet Cullet Specifications

Two methods of processing MSW glass to serve as an acceptable cullet are currently receiving much attention. They are froth flotation and optical sorting.

a. Froth Flotation

Froth flotation depends upon the tendency of hydrophobic particles to accumulate at the air/water interface of an aqueous system. To accomplish flotation, a cationic fatty compound (e.g., fatty amine) is added to an aqueous mixture of glass and nonglass particles. The fatty compound selectively adsorbs to the glass particles and thereby renders the particles hydrophobic. When air is blown through the mixture, the treated glass rises with the resulting air bubbles to form a froth at the surface of the mixture. The froth, with its glass content, is skimmed off the top of the water. The tailings, which are primarily nonglass particles, remain as a sediment at the bottom of the containing vessel. Removal of the glass from the froth and processing it into a suitable cullet

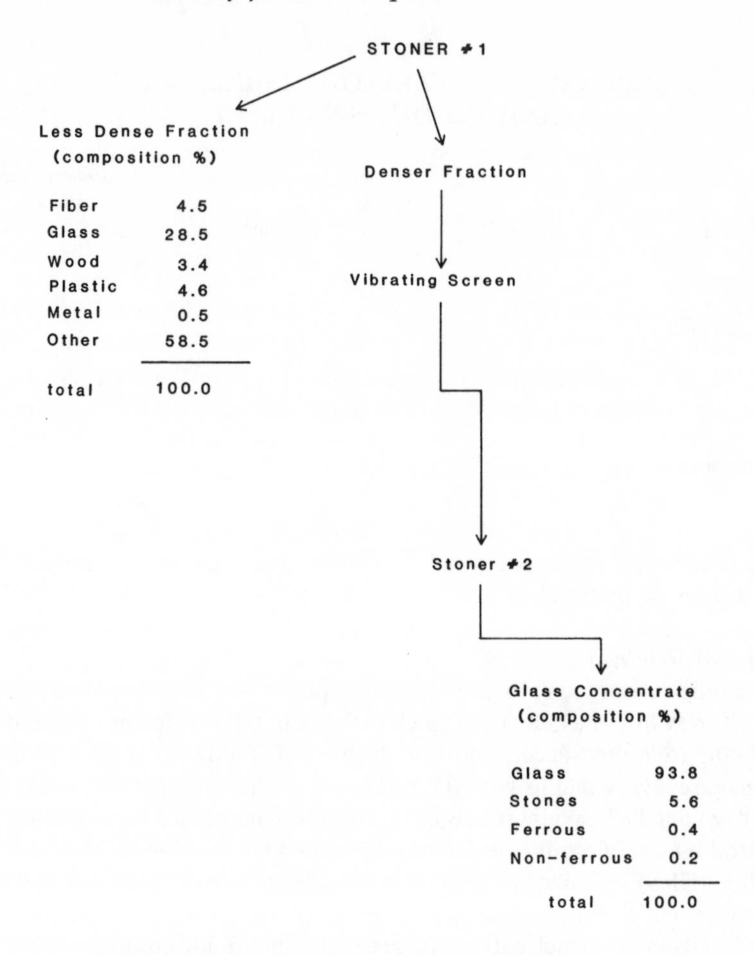

FIGURE 4. Flow sheet for the CRS glass concentration system.

requires several steps. In a system proposed and tested by the National Center for Resource Recovery (NCRR), the glass-rich slurry is processed through a series of three flotation steps in which the froth product from one flotation cell is introduced into the next cell. The froth product from the third flotation cell is dewatered in a spiral classifier and then dried in a rotary dryer to less than 0.5% moisture.

According to an *NCRR Bulletin*,[15] cullet produced by their version of the flotation process has a very low concentration of refractory particles, and glass produced from it is virtually free of inclusions. The cullet consists of a mixture of colors, and hence is suited only for container manufacture in which color is not a decisive factor.

b. Optical Color Sorting

In optical sorting, photocells are used to measure the intensity of light transmission through individual glass and nonglass particles and provide the impulses needed to trigger the mechanical separation of one class of particles from another. As of this writing, all available optical sorting systems were beset with shortcomings that rendered them unsuited for full-scale practical operations. Principal problems are slowness of throughput rates, incomplete recovery of glass, and incomplete separation of colors.

D. Economics of Glass Recovery

Inasmuch as mechanical glass recovery has as yet not been practiced in a full-scale resource recovery plant, no hard and fast figures can be given regarding the costs involved. While it is true that some numbers are available with respect to the economics of certain uses (also in the previous section "Uses for Recovered Glass"), it is a fact that unknown factors such as transportation and the doubtful nature of figures for capital and processing costs and market values, lend an air of uncertainty to the economics of glass recovery at present.

According to numbers cited by Duckett,[15] if a froth-separation facility were an "add-on" to a 546 to 910 Mg/day (600 to 1000 TPD) MSW plant, the cost of glass recovery (i.e., in 1978) would be on the order of $14/Mg of recovered cullet, exclusive of transportation. Color separation would add to the costs. With a going price of cullet at $27/Mg, Duckett saw little incentive for recovering glass from MSW until the development of a more favorable market.

IV. ALUMINUM

The aluminum industry not only is the youngest of the major nonferrous industries, it also has had the fastest growth. Thus, the U.S. primary aluminum industry grew from a single firm in 1939 to 12 major producers in 1978. The corresponding rise in production was from 148,000 Mg in 1939 to 5.2 million Mg in 1975.[29] Three factors threaten the continuation of this prodigious rate of growth, namely, the increasing scarcity of readily processed ore, its high price, and the cost of the energy required to process the ore. A consequence of the scarcity and high costs is a sharpened interest in recycling the aluminum in discarded products (i.e., secondary aluminum). The advantage or even necessity of recycling aluminum becomes obvious, when one considers that it takes only 5% as much energy to recover a kilogram of aluminum from scrap as it does to produce it from ore.[30,31]

The present section is restricted almost entirely to the recovery of aluminum from municipal solid waste (MSW), even though MSW is but one of several sources of the metal. The restriction is in keeping with the subject matter of this book, i.e., the recovery of resources from MSW. Aluminum is the principal nonferrous metal encountered in MSW, accounting as it does in some cases for as much as 90% of the total nonmagnetic metal fraction. However, in relation to the total solid waste stream, the amount seems less impressive in that it constitutes only from 1.0% to slightly more than 1.5% by weight of the total municipal waste stream. Nevertheless, because of the essential role had by aluminum in modern technology, the smallness of the fraction does not lessen the necessity nor reduce the energetic and economic advantages of recovering the metal from municipal waste.

A. Classification of Aluminum Waste

Despite the multitude of aluminum wastes, in terms of recycling they can be grouped into two general classes, namely, "wrought" and "cast". The division is a sharp one in terms of recycling because of the basic differences between the two classes with respect to chemical composition and metallurgy. The practical consequence of these differences is that the aluminum from cast alloys is not suitable for wrought alloys. On the other hand, while aluminum from wrought alloys can be used to produce cast alloys, the economics of so doing would be adverse. The reason for the latter is that generally a high quality wrought scrap is monetarily worth more than a casting scrap. Beverage containers and rigid foil are the more important sources of aluminum for wrought alloys, and for cast alloys, a broad assortment of household utensils and cooking ware. The diversity of chemical content between the many alloys is indicated by the data in Table 5.[8]

Scrap aluminum destined for reuse must be in a form with a low surface-to-volume ratio, and therefore must be either baled, briquetted, or shredded. Moreover, the amount of fines (aluminum, under 12 mesh), organic contaminants, sand, grit, and glass must be kept at a minimum. Specifications for baled density are established by the buyer and may range from 48 to 224 kg/m.[3,8] Usually, specifications call for fines not to exceed 3%, and free organics, about 2%. Examples of the target specifications for chemical analysis of the scrap are listed in Table 6. The subheadings (columns 1 through 3) refer to different grades of wrought scrap. In practice, column 1 deals with a wrought product, column 2 with a slightly off-grade wrought product, and column 3 with one of a poorer quality than that of column 2. The practical significance of the lower grades is in the need to increase the extent of dilution with primary metal as the quality declines. Moreover, the selling price of metal meeting the specifications in column 3 probably would be about one third less than that of the No. 1 grade.

B. Technology of Aluminum Recovery

The technology of aluminum recovery from the solid waste stream ranges from the extremely simple to the very complex. This spectrum is covered by household separation ("source separation") at one extreme to eddy current classification followed by air knife cleanup at the other extreme. The mechanics of source separation of aluminum items are so obvious as to warrant little or no explanation, whereas a dissertation on the political and social aspects of its implementation and on its advantages and disadvantages is beyond the scope of this book. Therefore, the remainder of this section is centered on mechanical processing for aluminum recovery.

1. Preprocess Steps

Almost without exception, existing mechanized aluminum removal systems require certain preparatory steps prior to the exposure of the waste stream to the separation device. The preparatory steps in effect accomplish a certain amount of classification and concentration of the aluminum fraction of the incoming wastes. The first step in practically all systems is size reduction. The size-reduced material is subjected to several unit processes, which among other things results in a nonferrous concentrate, of which aluminum is the major metallic constituent. Other constituents are heavy organics.

One of the unit processes, and an exceedingly important one, is the air classification step. In general, most of the aluminum material reports to the "heavies". However, difficulties do arise, primarily because aluminum is a light material. Moreover, the shapes of some of the aluminum in the waste stream are such as to impart airfoil characteristics with a consequent tendency to fly. As a result a trade-off must be made in the operation of the air classifier, namely, between a higher but "dirtier" yield of aluminum and a lower, but cleaner yield. The imposition of high air velocities results in a more efficient removal of lights and a reduction in the organic content of the heavies. Unfortunately, the high air velocities also lift more aluminum which then reports to the lights and, consequently, does not go on to the aluminum processing equipment. In fact, as much as 50% of the input aluminum can be lost by way of the lights. While it is true that an elaborate subsystem could be installed to remove aluminum from the lights, the costs and energy expenditure involved could make such a step impractical.

Ranking in importance with size reduction and air classification in preprocessing is the removal of magnetic materials. The importance of removing magnetics stems from the fact that they interfere with the functioning of aluminum separation equipment, and

Table 5
CHEMICAL ANALYSES OF COMMONLY USED ALUMINUM ALLOYS IN CONSUMER PRODUCTS

Alloy	Si	Fe	Cu	Mn	Mg	Zn	Ti	Others		Use
								Each	Total	
1235[a]	.65Si + Fe	—	.05	.05	.05	.10	.03	.03	—	Printed foil
3004	.30	.7	.25	1.0—1.5	.8—1.3	.25	—	.05	.15	Can bodies
5182[b]	.20	.35	.15	.20—.50	4.0—5.0	.25	.10	.05	.15	Can ends
8079	.05—.30	.7—1.3	.05	—	—	.10	—	.05	.15	Household foil
3003	.6	.7	.05—.20	1.0—1.5	—	.10	.10	.05	.15	Frozen food dishes
3105[c]	.6	.7	.30	.30—.8	.20—.8	.40	.10	.05	.15	Lawn chairs siding
6063[b]	.20—.6	.35	.10	.10	.45—.9	.10	.10	.05	.15	Extrusions
380[d]	7.5—9.5	1.3	3.0—4.0	.50	.10	1.0	—	—	.50	Die castings
2036[b]	.50	.50	2.2—3.0	.10—.40	.30—.6	.25	.15	.05	.15	Auto body sheet
5657	.08	.10	.10	.03	.6—1.0	.03	.10	.02	.05	Bright auto tri
7016	.10	.10	.6—1.4	.03	.8—1.4	4.0—5.0	.03	.03	.10	Bumpers
390	16.0—18.0	1.3	4.0—5.0	.10	.45—.65	.10	.20	.10	.20	Engine blocks

Note: Table uses Aluminum Association Limits, which are shown as either a range or a maximum.

[a] 99.35% Min aluminum required.
[b] Also allows .10 max Cr.
[c] Also allows .20 max Cr.
[d] Most commonly used die casting alloy—variations of this alloy allow up to 2% Fe, up to 3% Zn, and up to .35% Sn.

Table 6
CHEMICAL COMPOSITION SPECIFICATIONS FOR REUSE IN ALLOY PRODUCTS

	Wrought products, max weight			Cast and secondary alloy products, maximum weight
	No. 1 (%)	No. 2 (%)	No. 3 (%)	(%)
Si	0.30	0.65	1.00	1.00
Fe	0.70	0.85	1.00	1.00
Cu	0.25	0.65	1.00	2.00
Mn	1.50	1.50	1.50	1.50
Mg	2.00	2.00	2.00	2.00
Cr	0.10	0.10	0.30	0.30
Ni	0.05	0.10	0.30	—
Zn	0.25	0.65	1.00	2.00
Ti	0.05	0.05	0.05	—
Bi	0.02	0.03	0.30	—
Pb	0.02	0.03	0.30	0.30
Sn	0.02	0.03	0.30	0.50
Others, each	0.04	0.04	0.05	
Others, total	0.12	0.15	0.15	0.12
Aluminum	remainder	remainder	remainder	remainder

Modified from Abert, J. G., article reprint from *N.C.C.R. Bull.*, 7(2,3), 1977.

that ferrous metals constitute a contaminant in terms of final (aluminum) product specifications. Generally, the requisite freedom from magnetic contaminants is ensured by the institution of several stages of magnetic separation.

The sequence of typical preprocessing steps is indicated in the flow diagram in Figure 5. The concentrate produced in the preprocessing steps has the following characteristics: (1) its weight and its configuration are such that it is a part of the heavies in air classification; (2) it is almost completely free of magnetics; (3) it will include some rigid foil containers and balled household foil; and (4) it will not include much light aluminum foil.[32] It will not include aluminum ends of intact bimetal cans. Bimetal cans constitute a problem in metal recovery. The aluminum ends are lost to the aluminum stream because the cans become a part of the magnetic fraction. On the other hand, the aluminum constitutes a contaminant when the cans serve as a feedstock in the recovery of ferrous metal.

2. Separation of Aluminum

As with mechanical separation in general, processes for separating aluminum can be grouped into broad classes, namely, wet and dry separation. Examples of the dry processes are the Al Mag® separator, the Recyc-Al®, the Pulse-Sort®, and the "inclined ramp". The Al Mag® is usually accompanied by a device, the Air Knife®, to "clean-up" the output from the Al Mag®. Wet processes include the Black Clawson process and the heavy media processes.

a. Al Mag®

The trademark Al Mag® is an abbreviation of "aluminum magnet", a designation used by its manufacturer (Combustion Power Co.) when referring to the unit. Despite

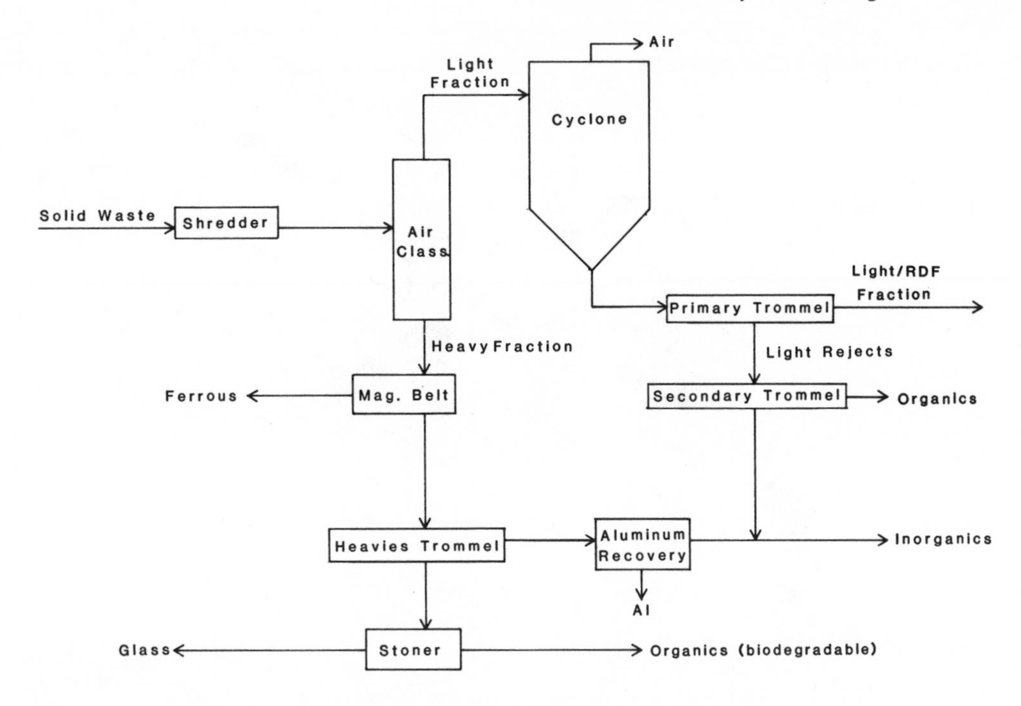

FIGURE 5. Position of aluminum recovery in Cal Recovery Systems flow plan.

the implication of the term "magnet", the functioning of the unit depends upon repulsion rather than attraction. The principle upon which the Al Mag® system is based is that metals passing through an electromagnetic field generate an electric current ("eddy current") in each piece of metal. The eddy current has a magnetic moment phased to repel the magnetic moment of the applied field. The repulsion results in the expulsion of nonferrous material from the waste stream exposed to the system. The primary objective in the use of the system is the recovery of aluminum cans, a product which would meet the specifications for a high-grade wrought.

The Al Mag® consists of four pairs of water-cooled electromagnets arranged about a 60-in.-wide conveyor in the manner indicated in Figure 6. The magnets utilize three-phase, 460-V power. Power consumption amounts to about 22 kW. Cooling the electromagnets involves a water flow of about 600 ℓ/hr. At a feed rate (particle size, 5 to 10 cm) of 0.063 Mg/hr, the removal efficiency may be as much as 99%. At a feed rate of 1.5 Mg/hr, it drops to 66%. The design rate of the machine is about 0.8 Mg/hr with an input material consisting of approximately 40% nonferrous material.

The Al Mag® leaves much to be desired in terms of its intended use, namely, the recovery of high-grade wrought. A major difficulty is that its sensitivity is not sufficient to permit a satisfactory differentiation between aluminum cans and other nonferrous items and shapes. Thus, the eddy current generated in a smaller but denser piece of cast aluminum may be the same as that arising from the movement of a larger but less dense piece of wrought in the field. In such a case, the repelling force either can be made to have a magnitude great enough to move both items, or small enough not to affect them, but it cannot be such as to move one but not the other. A second difficulty is that certain organics (e.g., textiles) entrained or trapped by the aluminum particles are removed with the aluminum. In test runs with the device, separated product was 20 to 25% in the form of free organics.[32]

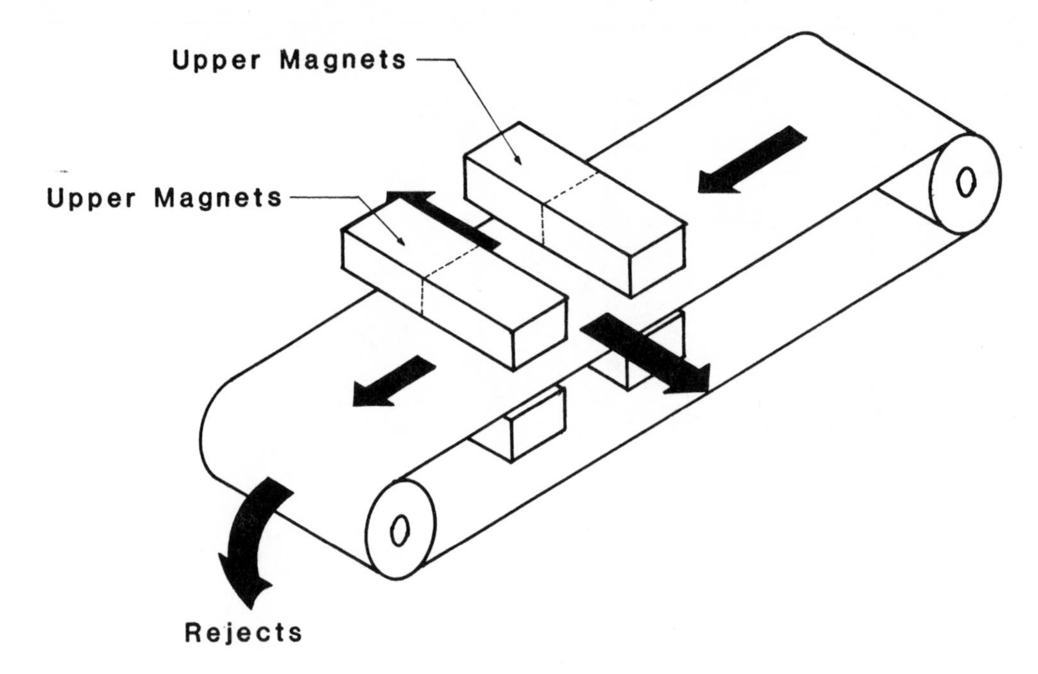

FIGURE 6. Diagrammatic sketch of "Al Mag".

A third difficulty is the inclusion of nonferrous (e.g., Cu and Zn) material other than aluminum.

b. Air Knife®

To lessen the extent of the entrainment of organics and of the inclusion of nonferrous material other than aluminum, the Aluminum Company of America developed a laboratory scale two-stage Air Knife.® The Air Knife® works on the principle of relative density. In the first stage it blows off the light organics, and in the second stage it blows the light aluminum can stock away from the heavier and generally more compact shapes of the red metals. Larger versions of the Air Knife® have been built recently, and are as yet in the testing stage of their development.

c. Recyc-Al®

The Recyc-Al® is another eddy current system (Occidental Research Corporation). The stainless steel belt used in the early versions of the Recyc-Al® system led to problems with tramp materials. Later is was found that substitution with a nonmetallic belt greatly mitigated the tramp material problem, and later versions accordingly were equipped with nonmetallic belts. Typically, the system involves a two-stage separation. Passage through the first stage results in a concentrate that is on the order of 80% metallic. The nonmetal content is further reduced to less than 5% in the second passage. Nonferrous metals not recovered in the Recyc-Al® system are poorly conductive alloys, nonmagnetic stainless steel, tangled wires, and foil-backed paper. The overall recovery efficiency is expected to be about 80% of the metals in the feedstream at its introduction into the system. The degree to which the mixed metal product from the system approaches the wrought specification is principally a function of the number of aluminum cans in the waste delivered to the resource recovery plant.

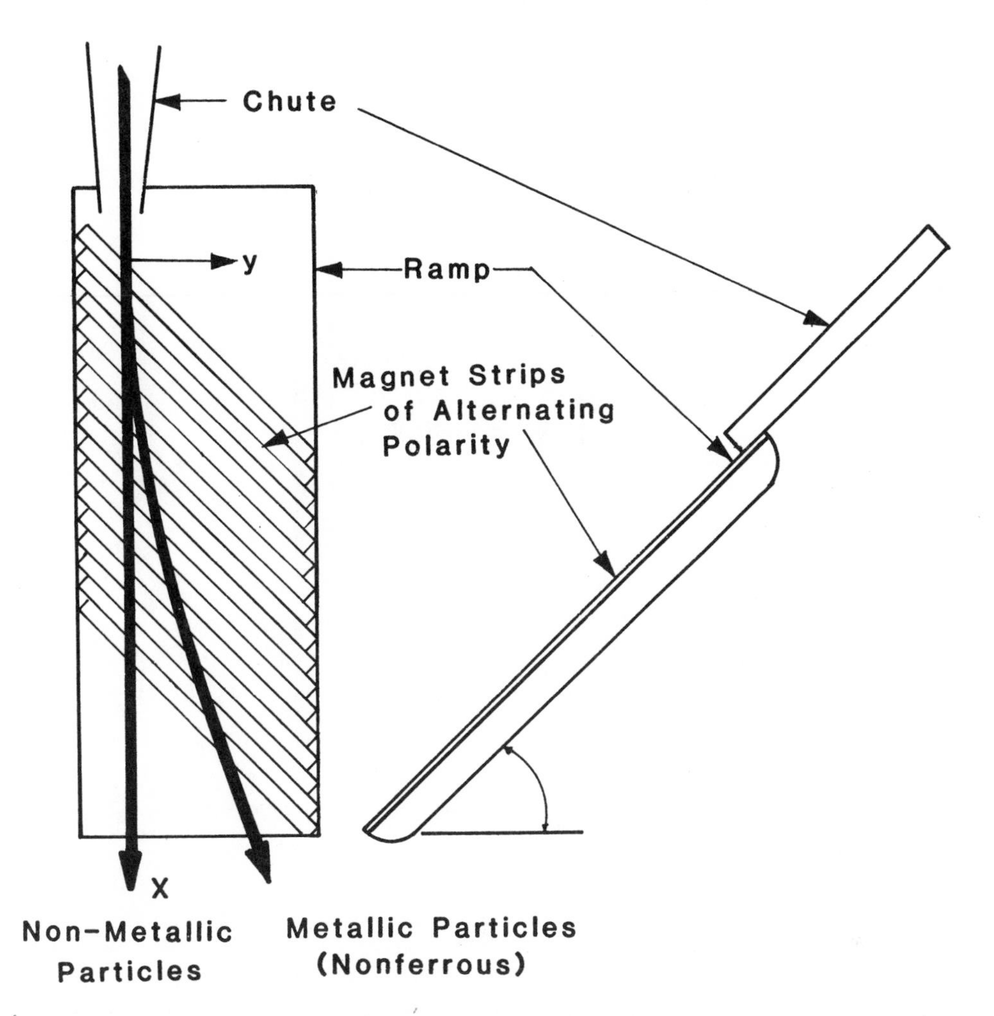

FIGURE 7. Schematic diagram of the inclined ramp nonferrous metal separator. (From Spencer, D. B. and Schiomann, E., *Res. Rec. Conser.*, 1, 153, 1975. With permission.)

d. Inclined Ramp

The inclined ramp system consists of inclined (45°) ramp in the surface of which are embedded permanent magnet strips of alternating polarity. The strips are inclined at an angle of approximately 45° to the ramp axis. The set-up is diagrammed in Figure 7.

In its operation, a nonferrous concentrate is introduced at a top corner of the ramp ("chute" in Figure 7) and is allowed to slide down. Not being affected by the magnetic field at the surface of the ramp, organic particles descend directly down the incline. On the other hand, an eddy current develops in the metallic particles as they move through the magnetic field. The current acts as a force with direction perpendicular to the magnetic strips. Due to the "force", the particles are deflected from a straight downward slide, and a separation is effected. The ability of the conductor to sustain the flow of the eddy currents is the selection criterion for identification and recovery of products. The repulsive force established by eddy currents is a function of particle size, geometry, and conductivity together with magnetic field strength and frequency. The material constant upon which the separation is based is the ratio of electrical

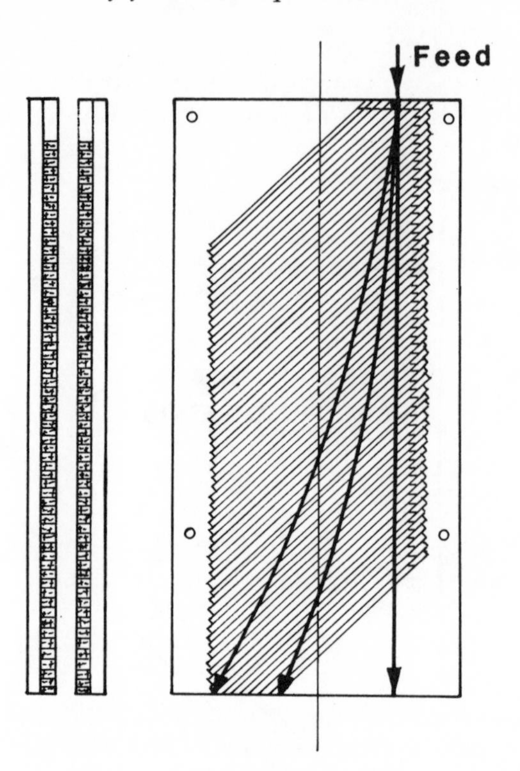

FIGURE 8. Vertical symmetric-field separator.

conductivity to density.[33] According to the developers of the device, from 70 to 80% of nonferrous metals in a chute feedstream can be recovered, and the metallic metals' purity of the product will be 90 to 98%. Two sequential ramps would be needed to accomplish such a separation. In a proposed system, a further separation into two products is accomplished, namely, a fraction that is mostly light aluminum and a second fraction consisting of heavy aluminum and the remaining nonmagnetic metals. This final separation would be accomplished with the use of heavy media.

e. Vertical Symmetric-Field Separator

The vertical symmetric-field separator[33] is based upon the free fall of particles in a symmetric magnetic field. A unit incorporating the system would consist of two vertical steel strips provided with magnetic strips and mounted under 45°, much in the manner shown for the "inclined ramp" in Figure 7. The arrangement of the two plates is indicated in Figure 8. An idea of the extent of the deflection brought about during the fall may be gained from the data in Table 7. The data were obtained in an experiment with the use of a prototype unit. The functioning of the system is not affected by particle shape excepting in extreme cases where aerodynamic frictional resistance reaches a significant level. An example of such a material is aluminum foil.

f. Heavy or Dense Media Separation

Heavy or dense media separation depends upon differences among metals in terms of their densities and upon the use of media that have a specific gravity greater than 2.6. Aluminum floats on such media, whereas stainless and the red metals sink. The medium generally is finely ground ferrosilicon or magnetite suspended in water. The density of the liquid is determined by its concentration of finely ground material. A heavy medium

Table 7
DEFLECTION OF NONFERROUS METALS IN A VERTICAL SYMMETRIC FIELD SEPARATOR

Material	Deflection (mm)
Brass	
Spheres	93
Cylinders	90
Copper	
Spheres	289
Cylinders	269
Aluminum	
Spheres	445
Cylinders	434

Modified from Dalmijn, W. L., Voskuyl, W. P. H., and Roorda, H. J., *Recycling Berlin '79,* Proc. Int. Recycling Congr., Thome-Kozmiensky, K. J., Ed., Berlin, 1979, 930.

process commercially available at the time of this writing is the Dutch State Mines (DSM) process.[34]

For the present, the costs involved in setting up and operating a heavy media system are so great as to preclude its use for recovering aluminum from municipal wastes. However, its use is not uncommon in automobile shredding operations.

g. *Black Clawson Method*

The Black Clawson system involves wet processing for resource recovery, in that in its operation all incoming refuse is slurried. The slurried refuse is processed in a series of steps. The portion of the process concerned with aluminum recovery is indicated in the flow diagram shown in Figure 9. As indicated in the figure, the first step in aluminum recovery begins directly after the initial hydropulping when the slurry is passed through the cyclone. The cyclone separates the heavier constituents of the slurry from the lighter constituents. The aluminum is in the heavier fraction. The second step takes place after the heavy material has been successively screened, exposed to a magnet, and placed in a heavy dense medium (see Figure 9). The fraction that sinks in the heavy medium is a concentrate consisting of glass and aluminum and other nonferrous metals, and an assortment of inert materials (e.g., rocks). The aluminum and nonferrous concentrate is subjected to jigging to separate the light fraction from the heavy fraction. Separation is accomplished by pulsating the water in the jig. Light materials are carried to the top, whereas heavier particles remain near the bottom. In reality, the separation is threefold: tailings, "light cup", and "heavy cup". The "light cup" is a glass concentrate that includes some aluminum and other nonferrous metals. The "heavy cup" includes cast aluminum and heavy nonferrous metals.

The aluminum in the glass concentrate is recovered electrostatically. In order to do so, the concentrate must be dried. Separation of glass from metals depends upon differences in conductivity between the two. Metals thus separated are added to the "heavy cup" product.

C. Status of Mechanized Aluminum Recovery

As of this writing, no full-scale operation is functioning successfully. In fact, almost all experience has been with small-scale, well-controlled operations. As a consequence,

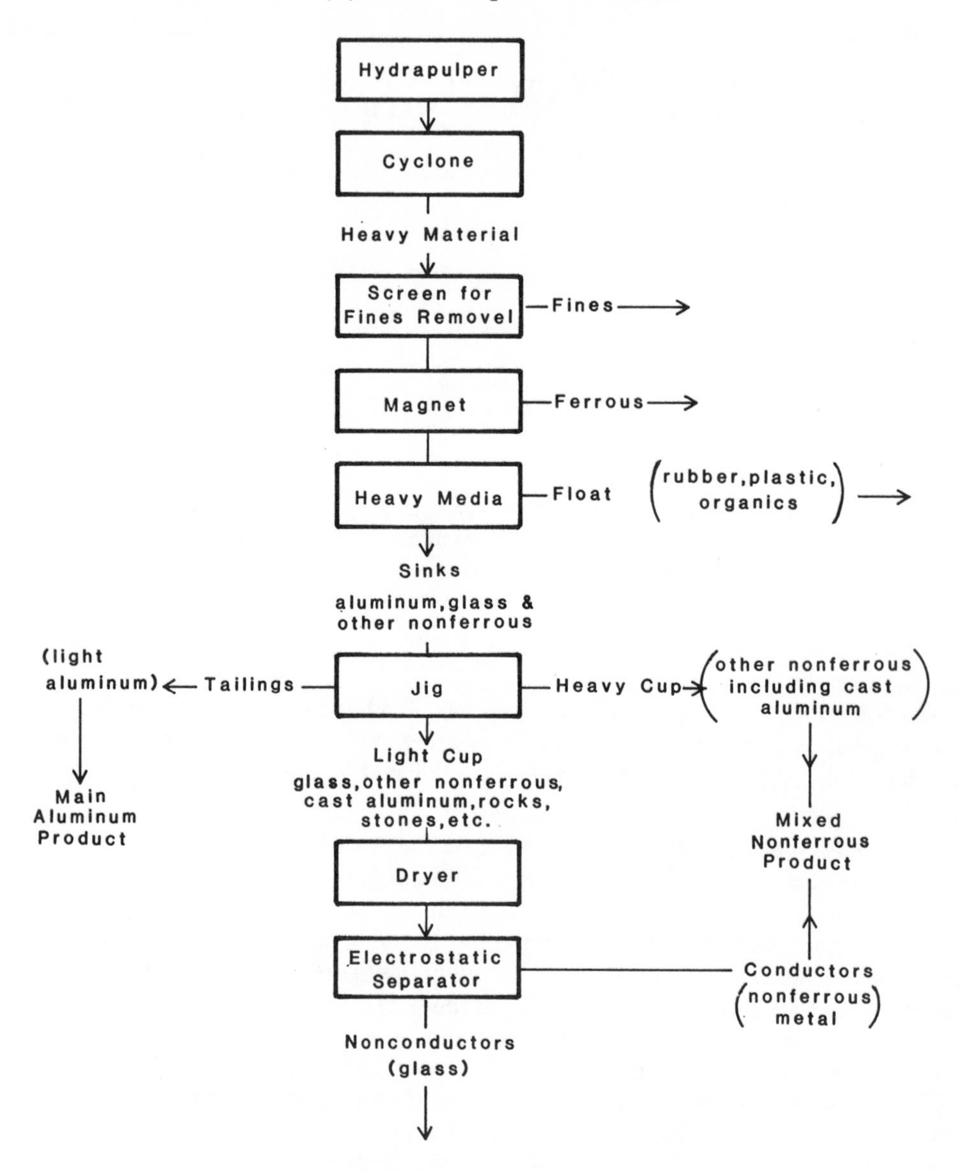

FIGURE 9. Points of aluminum recovery in the Black Clawson system.

the values for efficiencies and other performance parameters quoted herein must be treated with a great deal of uncertainty when making projections for design or other purposes.

A major problem characteristic of existing aluminum separation systems is the low rate of throughput that must be applied to ensure a satisfactory product. This throughput limitation coupled with substantial initial costs renders the operation marginally economically feasible at best. The low throughput rate combined with a less than satisfactory separation makes it apparent that much remains to be done in the improvement of the aluminum separation technology.

V. MAGNETIC METALS (FERROUS METALS)

The term "magnetic solid waste" generally is applied to the discarded iron and steel materials and magnetic components of discarded products that are not separately collected and processed into iron and steel by the ferrous scrap industry. In practice, magnetic waste materials are referred to as "iron and steel scrap" only after they have been processed into a form that meets the specifications laid down by the various components of the iron and steel industry. The concentration of magnetic metals in municipal solid waste varies from about 3 to 8% by weight, and is dependent upon the particular area of collection. In terms of tonnage, the amount of magnetic metals discarded each year in the 1970s as residential and commercial waste amounted to 10 million Mg.[35] This yearly amount is much less than that of the ferrous purchased by the nation's steel mill. For instance, in 1977 the steel mills and foundries purchased 38.5 million Mg of ferrous scrap for recycling.[36] The larger amount is a result of the fact that iron and steel scrap constitute about 50% of the total feedstock to the U.S. iron and steel industry, and iron ore, the remaining 50%. Moreover, the use of the steel scrap resulted in an energy savings of about 7.7×10^{14} kJ (7.3×10^{14} Btu).

The three major types or classes of iron and steel scrap are "home scrap", "prompt industrial scrap", and "obsolete scrap". The positions of the three types in the iron and steel cycle are shown in the flow diagram in Figure 10.

"Home scrap" is generated by steel companies and foundries in steel-making and finishing operations. The material typically is of good quality and has a known composition. Consequently, its usage involves a minimum of processing, and normally it is recycled within the plant in which it was generated. "Home scrap" accounts for approximately 60% of the nation's total output of ferrous scrap.

"Prompt industrial scrap" is the ferrous scrap produced in the metal working industry. It consists of chips, "turnings", and "punchings". Normally, the greater part of the prompt industrial scrap is recycled immediately and directly because it is relatively clean and readily accessible. "Prompt industrial scrap" accounts for about 20 to 25% of the total scrap production.

Magnetic materials discarded after consumer use collectively bear the designation "obsolete scrap". Common examples of "obsolete scrap" are furniture, containers, and appliances. "Obsolete scrap" constitutes 15 to 20% of the total U.S. output of scrap. "Prompt industrial scrap" and "obsolete scrap" also are referred to collectively as "purchased scrap", because typically they are handled by a scrap dealer. The major types of iron and steel scrap are subdivided into 42 grades of iron and steel scrap, 29 grades of alloy steel scrap, and 40 grades of railroad scrap.

The scrap best suited to the needs of the iron and steel industries is one that has these three characteristics: a known composition, a minimum concentration of contaminants, and generation from a known source. Because it has the preceding three characteristics, "home scrap" is regarded as being the most desirable of the three main classes of scrap. It is agreed that the physical characteristics of the "prompt scrap" generated in some areas of the industry can be upgraded through the imposition of relatively strict segregation procedures. While it may be lacking in the qualities that make "home" and "prompt scrap" preferable, the obsolete scrap found in municipal solid waste (MSW) does provide the greatest potential for increasing the current level of magnetic metal recycling. For that reason, as well as because the principal concern of this book is with urban wastes, the presentation that follows is focused on "obsolete scrap", i.e., magnetic solid waste.

A. Quantity and Composition of Magnetic Solid Waste

The amount of magnetic metals found in MSW varies not only from nation to nation, but also from region to region and even city to city. The factors responsible for the variation are the same as those that determine the characteristics of the wastes from a given source. Among the more important of the factors are degree of industrialization, social economic development, and climate. For example, the MSW generated in India

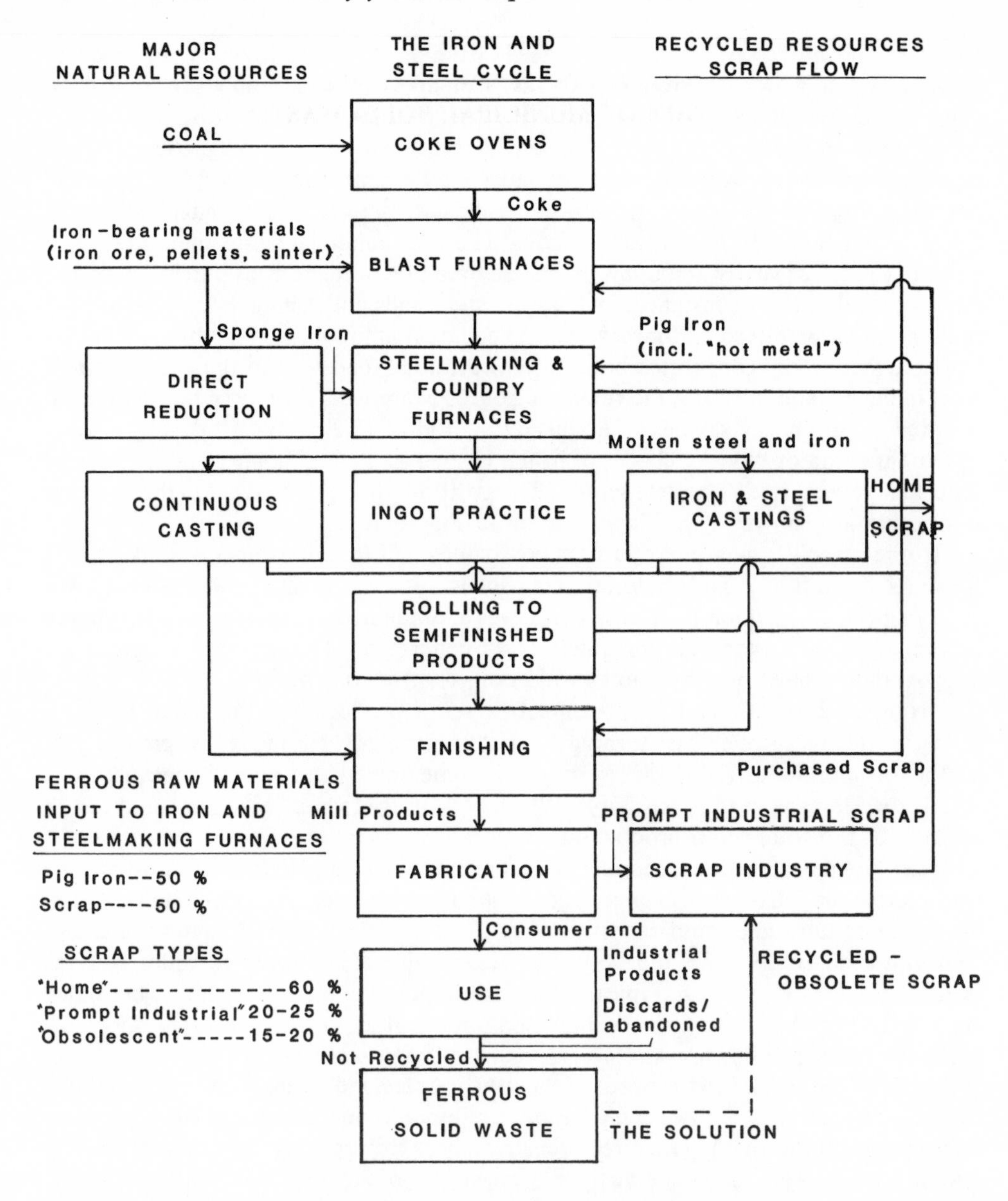

FIGURE 10. The iron and steel cycle and ferrous solid waste.

contains less than 1% magnetic metals, whereas as much as from 6 to 8% of the MSW generated in Australia and in the U.S. may be in the magnetic form.[8,37]

The average composition of the magnetic fraction of MSW in the U.S. is indicated by the data in Table 8. According to the table, steel containers account for 50 to 60% of the magnetic fraction; appliances, 16%; and a miscellany of items, 24 to 36%. Of the containers, those for food constitute the largest share. As much as 90% of the magnetic fraction of residential wastes may be in the form of steel cans.[41]

Inasmuch as steel cans make up a sizeable portion of the magnetic waste stream, a few words about them are in order at this point. According to the Can Manufacturers Institute, since 1970 the yearly average production of cans in the U.S. since 1970 has been in excess of 65 billion units. More than 5 million Mg of steel were used in the production of the 65 billion units.

Table 8
MAJOR COMPONENTS OF THE MAGNETIC METAL FRACTION OF MUNICIPAL SOLID WASTE[a]

Item	Weight (%)
Cans	
Food	23.5—28.2
Beer	11.5—13.8
Soft drinks	8.0—9.6
Other	7.0—8.4
Total	50—60
Appliances	16
Miscellaneous items	24—36

[a] Based on work by Regan,[38] Hill,[39] and Garbe.[40]

Table 9
COMPOSITION OF THE THREE TYPES OF STEEL CONTAINERS

Component	Tin-free steel aluminum end (%)	Tin-plated aluminum end (%)	Steel tin-steel end (%)
Organic coating	1.83	1.83	1.83
Aluminum	10.20	10.20	0.00
Lead	0.01	1.83	1.84
Tin	0.00	0.42	0.48
Steel	87.96	85.71	95.85
Total	100.00	100.00	100.00

Modified from Alter, H. and Reeves, W. R., Specifications for Material Recovered from Municipal Refuse, EPA-670/2-75-034, May 1975.

Steel containers are classified into three basic groups, namely, tin-lined cans (the so-called "tin cans"), bimetallic cans, and tin-free cans. The tin lining has a twofold purpose (1) it serves as a base for soldering the seams; and (2) it protects the steel against corrosion by acidic contents, thereby preventing the contamination of the container contents by ferrous compounds. The tin-free can was developed in response to the steep rise in the price of tin and to the fluctuation in the supply of metal. With the tin-free container, a combination of chromium and chromium oxide serves as a solid ring base, while special resins fill the protective function had by tin. The composition of the three types of steel containers is given in Table 9. With the exception of steel, each of the substances listed in Table 9 is a source of problems when used in the production of steel.

B. Methods of Recovery

In countries in which labor is expensive or in short supply, methods of recovering magnetic metals from the MSW stream are based in large part upon the magnetic permeability of ferrous metals. In comparison with other metals, iron has a very strong magnetic permeability. Ferromagnetic materials can be recovered from a mixture of

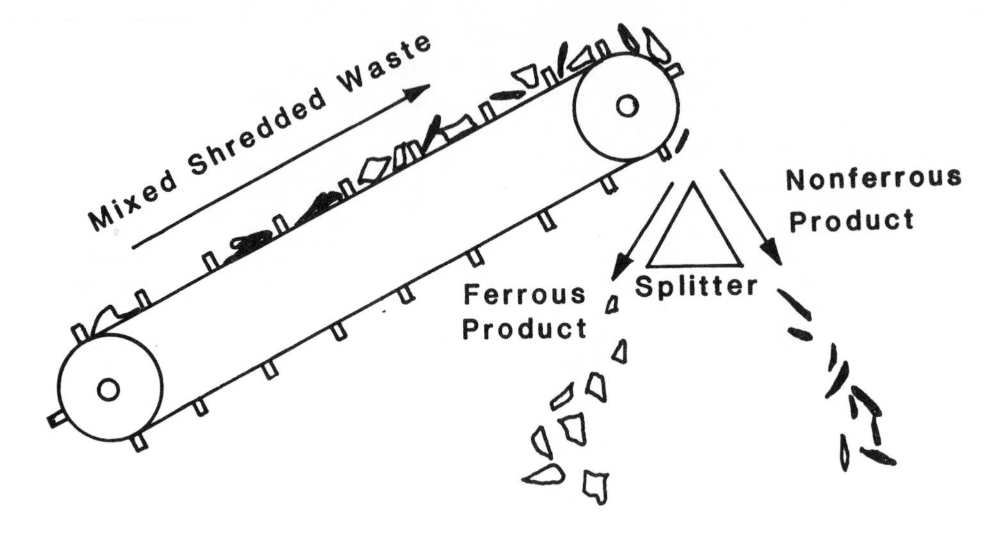

FIGURE 11. Simple magnetic head pulley conveyor.

ferrous and nonferrous materials by passing the mixture through a magnetic field. The field exerts magnetic forces which, coupled with other forces such as gravity and friction, cause the materials to follow different paths. The principles and technology of magnetic separation are fairly well developed because of extensive utilization of the process to beneficiate iron ores in the mineral processing industry.

1. Equipment

Magnets may be either of the permanent or the electromagnetic type. They come in one or more of three configurations, namely, the drum, the magnetic head pulley, and the magnetic belt. They may be assembled and suspended in line, crossbelt, or mounted as head pulleys in the material transfer conveyor as its name implies. The magnetic head pulley-conveyor consists of a magnetic pulley mounted in a conveyor. A schematic diagram of the configuration is shown in Figure 11. In its operation, the material to be sorted is passed over the pulley in such a manner that the nonferrous material will fall free by gravity vertically onto the next conveying device. In the suspended drum configuration, the electromagnetic assembly usually is mounted stationary inside an outer rotating drum, as is shown in Figure 12. The drum magnetic assembly can be installed for either overfeed or underfeed. The magnetic belt in its simplest form consists of simple, single magnets mounted between two pulleys which support the conveyor belt mechanism. The magnetic belt and its position in the processing line are diagrammed schematically in Figure 13.

The energy required to operate the magnets vary with the type of magnet and configuration of arrangement. Reported requirements range from 8 to 14 kW for the magnets. These numbers do not include energy required by other parts of the set-up.[44]

2. Efficiency Factors

The efficiency of magnetic metal recovery from shredded refuse in terms of weight of magnetic metal recovered per unit weight of magnetic metal in the in-feed stream typically is about 80%.[43] In the absence of definitive studies pointing to the contrary, it is safe to assume that the efficiency of ferrous metals recovery from the heavy fraction separated through air classification would be on the order of 85 to 90%.

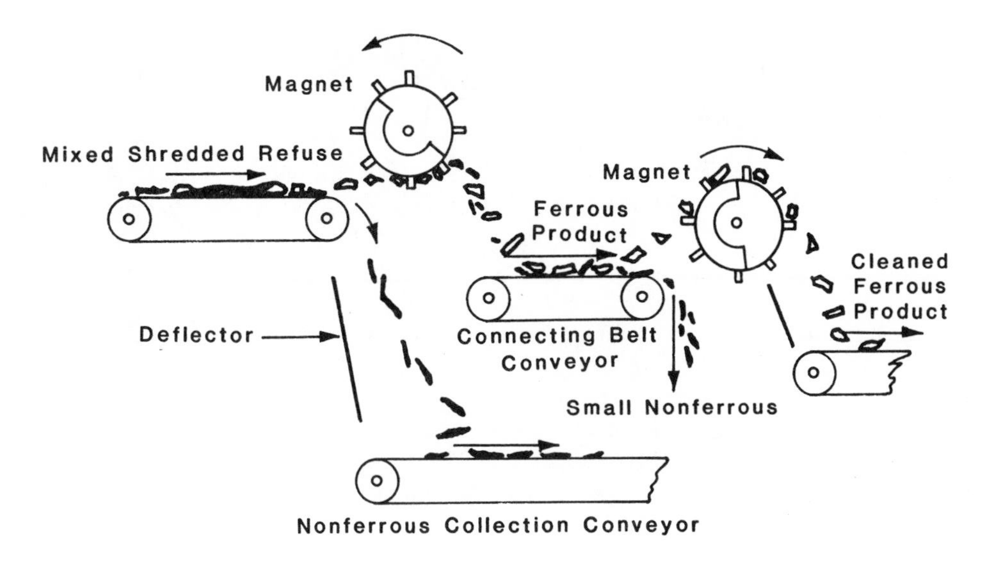

FIGURE 12. Multiple drum magnetic drum and conveyor.

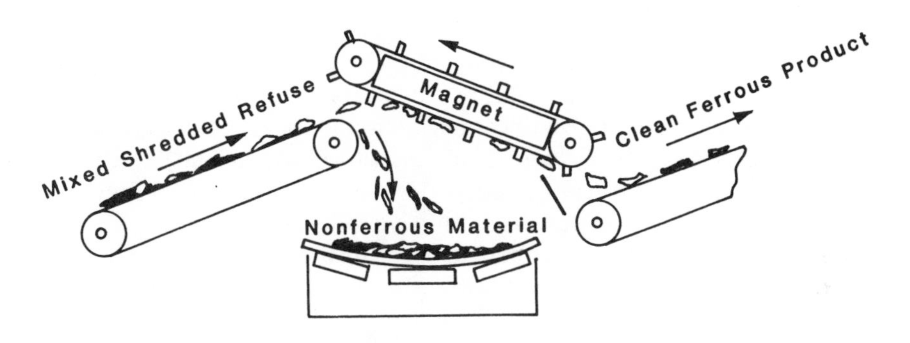

FIGURE 13. Magnetic belt conveyor.

Thus far, it has been the general experience that ferrous scrap recovered by a magnetic separator placed directly downstream of primary size reduction equipment is inferior in quality to that removed by a separator placed in an air-classified heavy stream. The reason for the difference in quality is that the air classifier effectively removes from the MSW most of the paper, plastic, rags, and other contaminants that otherwise might cling to or be entrapped by the ferrous scrap, or be ballistically carried over with the metal. On the other hand, no such cleanup takes place before the material is passed through the shredder, nor does any take place in the shredder. Therefore, contaminants left entrained by or adhering to the ferrous scrap are removed with it or are carried over ballistically.

The amount of "pickable" contaminants to be found in the scrap removed directly after the shredding step may range from 0.2 to 1.0% of the material removed by the magnetic separator. The contaminants are chiefly in the form of paper, plastic, and rags. Inasmuch as the contaminants are substances of a low density as compared to the high density of clean ferrous scrap, the relatively low (0.2 to 1.0%) contamination on a gravimetric basis assumes a much greater significance when translated into volumetric terms. The latter bespeaks a visually unpleasant product. The literature available at the time of this writing contains no reports regarding the degree of contamination of ferrous scrap removed from air-classified heavies. Nevertheless, it can be safely assumed that the amount would be only one third to one fifth that reported for shredded MSW not air classified.

Other factors that affect the degree of cleanliness of recovered ferrous scrap are the burden-depth of the material, speed of the conveyor belt, strength of the magnetic field, and magnet configuration. Of the four, the first three have the greatest bearing on the successful recovery of ferrous scrap. With a given magnetic separator and at a given belt speed, the quality of the recovered ferrous scrap improves as the layer of the burden becomes shallower. Lowering the depth of the burden lessens the chances of adhering or entrapped paper, plastic, and rags being removed simultaneously with the ferrous metallic items to which they are attached. As a consequence, the quality of the ferrous scrap is correspondingly improved.

At a given burden-depth, the yield of ferrous scrap increases with decrease in the size of the gap between magnet and conveyor belt. The increase is due to an augmentation in the field strength of the magnetic field brought about by the shortening of the distance between conveyor belt and magnet.

Velocities of conveyor belts as they pass the magnetic separator generally are in the range of 1.5 to 2.2 m/s. In practice, the possibility of reducing the required belt width generally plays a more important part in determining belt speed that does the bearing of belt speed on efficiency of removal.

C. Uses for Separated Magnetic Wastes

The two most common uses for ferrous metal separated from MSW are in steel production and in the precipitation of copper. Before the steel in containers can be put to either of these uses, it must be freed of nonferrous metals listed in Table 9, especially the tin and the aluminum. With tin removed, the steel in containers generally is of a high quality, and therefore is well suited to use in steel production. For transport to the steel mills, the detinned containers are compressed into bales at the detinning plant.

1. Detinning

Tin typically is removed from tin plate by immersing the plate in a hot solution of sodium hydroxide to convert the tin to sodium stannate. The sodium stannate solution is then subjected to electrolysis to return the tin to its elemental state. The tin thus recovered is melted and cast into ingots. Approximately 6 lb of tin can be recovered from every ton of tin cans processed. The effectiveness of the detinning process is adversely affected by the presence of nonmetallic material (e.g., food wastes, paper) trapped in or adhering to the containers. Aluminum "contaminants" are especially troublesome in the detinning process, because they react with the sodium hydroxide to form sodium aluminate. This reaction, in addition to using up the reagent, results in frothing and spilling of the caustic solution. To keep the amount of "contaminants" at a minimum, the detinning step is preceded by one or more of the following steps: size reduction and magnetic separation, air classification, and even combustion. If the detinned cans are intended for use as ferrous scrap, the combustion step is omitted because it results in the fusion of some tin to the steel and thereby lessens the utility of the scrap for steel production.

2. Copper Precipitation

Detinned steel is used in the extraction of copper from low-grade ore and mining wastes. In the extraction process, the ores and wastes, some of which may contain as little as 0.1% copper, are treated with a solution of sulfuric acid. The copper reacts with the acid to form copper sulfate. The resulting copper sulfate solution is allowed to leach through a bed of the detinned steel. During the course of the leaching copper is precipitated out of the solution through the following reaction:

$$Cu\,SO_4 + Fe \rightarrow Cu + FeSO_4$$

The precipitate is known as "cement copper". From 1.5 to 2.5 Mg of scrap are required in the production of one ton of copper.

Because of their cleanliness and somewhat uniform size, detinned scrap and residue from can manufacturing such as punchings and clippings are particularly suitable for copper precipitation.

Cans recovered from municipal solid waste are not in a condition suited for immediate use in a copper precipitation process. They must be shredded and be freed of contaminants. (The purpose of the shredding is to increase the ratio of surface area to mass.) Processing plants have been especially designed to accomplish these two essential steps by way of size reduction and incineration. The incineration step is included as a means of removing organic contaminants. It must be carried out under controlled conditions in order to prevent air pollution, excessive oxidation of the metals, and coating of the metal with glass. Coating of the magnetic scraps with glass takes place rather frequently. The coating seriously impedes the precipitation process.

3. Specifications

Specifications generally delineated by the detinning industry are

1. The scrap must be 95% magnetic metal and should be free of garbage, paper, and plastics.
2. Balled or "nuggeted" scrap is unacceptable.
3. The scrap should not include massive ferrous objects.

D. Cost of Magnetic Separation

The data presented in this section are taken mostly from writings by Alter and Woodruff.[44] According to Alter and Woodruff's estimate, in late 1976 the cost of a magnetic separator and all of its essential appurtenances was about $100,000 for a 45-Mg/hr (50 tons/hr) system. The amount includes supports and a hopper to receive and store the separated metal. To determine the actual installed cost of a complete system, it is necessary to have information as to the specifications for the recovered metal and the requirements specific to the site. Alter and Woodruff[44] arrive at what they state as being conservative, a cost estimate of $2,412,000 for a complete plant having a capacity of 45 Mg MSW/hr and covered storage for 450 Mg of MSW. The estimate includes the cost of a shredder but *not* of land and of an air classifier. According to a table presented in the reference, the total direct costs of the equipment amounted to $1,225,000, and of the building installation, $485,000. Indirect costs were estimated as being $702,000. They projected an operating cost of $993,000, or about $6.05 per input Mg based on processing 164,000 Mg MSW per year.

Tobert[45] estimates the capital, annual operating, and annual maintenance costs (in 1976) for a belt type magnet (45-Mg/hr feed) as being $32,000, $700, and $800, respectively. For the drum, his estimates are $42,000, $500, and $1000, respectively. On the basis of these estimates, the cost per ton of material processed would be $0.047 with the belt type magnet and $0.057 with the drum magnet.

REFERENCES

1. Statistical Abstracts of the Unted States, 99th annu. ed., U.S. Department of Commerce, Bureau of Census, Washington, D.C., 1978.
2. **Iannazzi, F. D. and Firth, L. M.,** Waste paper recovery comparison of U.S. rates with those of selected foreign countries, *Tappi,* 61(6), 23, 1978.
3. **Iannazzi, F. D.,** Comparison of fiber and energy values of waste paper, paper presented at TAPPI Pulping/Secondary Fibers Conf., Washington, D.C., November 1977.
4. **Lischer, V. C., Jr.,** Solid waste as supplementary boiler fuel for paper mills, *Tappi,* 59(6), 104, 1976.
5. **Bridgwater, A. V. and Mumford, C. J.,** *Waste Recycling and Pollution Control Handbook,* Van Nostrand Reinhold Environmental Engineering Series, Van Nostrand Reinhold Company, New York, 1979.
6. *Paper Stock Standards and Practices, Circular PS-74,* Paper Stock Institute of America, New York, 1974.
7. **Nollet, A. R. and Sherwin, E. T.,** Paper pulps from municipal solid waste, in *Proc. 1980 Natl. Waste Process. Conf., 9th Bienn. Conf.,* Washington, D.C., Am. Soc. Mech. Eng., New York, N.Y., May 11 to 14, 1980.
8. **Diaz, L. F., Savage, G. M., Goebel, R. P., Golueke, C. G., and Trezek, G. J.,** Market potential of materials and energy recovered from bay area solid wastes, report prepared for the State of California Solid Waste Management Board, March 1976.
9. **Trezek, G. J., and Golueke, C. G.,** Availability of cellulosic wastes for chemical or biochemical processing, *AIChE Symp. Ser.,* 72(158), 52, 1976.
10. **Savage, G. M., Diaz, L. F., and Trezek, G. J.,** Fiber from Urban Solid Waste Recovery Procedures and Pulp Characteristics, *Tappi,* 61(6), 6, 1978.
11. **Yoda, T., Miyazaki, T., and Machida, O.,** Making pulp from municipal refuse, in *Recycling Berlin '79,* Proc. Int. Recycling Congr., Thome-Kozmiensky, K. J., Ed., E. Freitag-Verlag fur Umwelttechnik, Berlin, 1979, 1176.
12. **Renard, M. L.,** Technological barriers to the recovery of secondary fibers from municipal solid waste, in *Proc. 1978 Pulping Conf.,* Tappi Press, Atlanta, 1978, 333.
13. **Renard, M. L.,** Selective dry processing, pulp cleaning and paper making tests on three feedstocks from municipal solid wastes, in *1979 Pulping Conf. Proceed.,* Tappi Press, Atlanta, 1979, 247.
14. Fourth Report to Congress—Resource Recovery and Waste Reduction, EPA Report SW-600, U.S. Environmental Protection Agency, Washington, D.C., 1977.
15. **Duckett, E. J.,** Glass recovery and reuse, *N.C.R.R. Bull.* 8(4), 87, 1978.
16. Baltimore tries squeezing out RDF profits, *Waste Age,* 11(5), 10, 1980.
17. ASTM, Standard methods of testing waste glass as a raw material for manufacture of glass containers, Designation: E688-79, American Society for Testing and Materials, in *Annual Book of ASTM Standard,* 1979, 952.
18. **Malisch, W. R., Day, D. E., and Wixson, B. G.,** Use of waste glass for urban paving, in *Proc. and Miner. Waste Utilization Symp.,* U.S. Bureau of Mines and IIT Research Institute, Chicago, 1972, 369.
19. **Connor, J.,** Marion County experiments with paving mix containing paving mix material from waste for paving roads, *Waste Age,* 2(5), 24, 1971.
20. **Malisch, W. R., Day, D. E., and Wixson, B. G.,** Use of salvaged glass in bituminous paving, in Proc. Spec. Centen. Symp., Technology for the Future to Control Industrial and Urban Wastes, University of Missouri, Rolla, 1971, 26.
21. **Day, D. E., Malisch, W. R., and Wixson, B. G.,** Improved bonding of waste glass aggregate with bituminous binders, *Am. Ceram. Soc. Bull.,* 40(12), 1038, 1970.
22. **Phillips, J. C. and Cahn, D. S.,** Refuse glass aggregate as an ingredient of lightweight aggregate, in *Proc. 3rd Miner. Waste Utilization Symp.,* U.S. Bureau of Mines and IIT Research Institute, Chicago, 1979, 385.
23. **Shotts, R. W.,** Waste glass as an ingredient of lightweight aggregate, in *Proc. 3rd Miner. Waste Utilization Symp.,* U.S. Bureau of Mines and IIT Research Institute, Chicago, 1972, 411.
24. **Willey, C. R. and Bassin, M.,** The Maryland environmental service/Baltimore County resource recovery facility, Texas, Maryland, *Proc. 6th Miner. Waste Utilization Symp.,* U.S. Bureau of Mines and IIT Research Institute, Chicago, 1978.
25. **Tyrell, M. E. and Goode, A. H.,** Waste Glass as a Flux for Brick Clays, Reports of Investigations 7701, Bureau of Mines, U.S. Department of Interior, Washington, D.C., 1972.
26. **Tyrell, M. E. and Feld, I. J.,** Fabrication and Cost Evaluation of Experimental Brick from Waste Glass, Report of Investigations 7605, Bureau of Mines, U.S. Department of Interior, Washington, D.C., 1972.

27. **Goode, A. H., Tyrrell, M. E., and Feld, I. J.,** Glass Wool from Waste Glass, Report of Investigations 7708, Bureau of Mines, U.S. Department of Interior, Washington, D.C., 1972.
28. **Lubisch, G.,** Energy savings by remelting of scrap glass in the container glass energy, in *Recycling Berlin '79, Proc. Int. Recycling Cong.,* Thome-Kozmiensky, K. J., Ed., E. Freitag-Verlag fur Umwelttechnik, Berlin, 1979, 1241.
29. *Annual Statistical Review,* Aluminum Association, 1975.
30. *Use of Aluminum in Automobiles: Effect on the Energy Dilemma,* Aluminum Association, 1977.
31. **Atkins, P. R.,** Recycling can cut energy demands dramatically, *Eng. Mining J.,* 174(5), 6971, 1973.
32. **Abert, J. G.,** Aluminum recovery—a status report, article reprint from *N.C.R.R. Bull.,* 7(2,3), 1977.
33. **Dalmijn, W. L., Voskuyl, W. P. H., and Roorda, H. J.,** Low-energy separation of nonferrous metals by eddy current techniques, in *Recycling Berlin '79, Proc. Int. Recycling Congr.,* Thome-Kozmiensky, K. J., Ed., Berlin, 1979, 930.
34. **Easterbrook, G. E.,** Aluminum can't resist the power of the medium, *Waste Age,* 10(1), 16, 1979.
35. Resource Recovery and Waste Reduction, fourth report to Congress, Environmental Publication SW-600, U.S. Environmental Protection Agency, Office of Solid Waste Management Programs, Washington, D.C., 1977, 142.
36. Institute of Scrap Iron and Steel, Inc., Recycling Iron and Steel Scrap Saves Energy, undated, p.16.
37. **Golueke, C. G. and Diaz, L. F.,** Organic wastes for fuel and fertilizer in developing countries, prepared for the United Nations Industrial Development Organization, 1980.
38. **Regan, W. J., James, R. W., and McLeer, T. J.,** Identification of opportunities for increased recycling of ferrous solid waste, prepared for the U.S. Environmental Protection Agency, NTIS Report PB-213-577, 1972, 33.
39. **Hill, G. A.,** Steel can study, prepared for the U.S. Environmental Protection Agency, Washington, D.C., 1973, 103.
40. **Garbe, Y. M. and Levy, S. J.,** Resource Recovery Plant Implementation: Guide for Municipal Officials—Markets, Report No. SW-157.3, U.S. Environmental Protection Agency, 1976, 47.
41. **Alter, H. and Crawford, B.,** Materials recovery processing research—a summary of investigations, report prepared for the U.S. EPA under contract No. 67-01-2944, October 1976, 101.
42. **Alter, H. and Reeves, W. R.,** Specifications for Materials Recovered from Municipal Refuse, EPA-670/2-75-034, May 1975, 109.
43. **Bendersky, D. and Simister, B.,** Study of Processing Equipment for Resource Recovery Systems: Volume II—Magnetic Separators, Air Classifiers, and Ambient Air Emission Tests, Midwest Research Institute, under contract to the U.S. EPA, 1979.
44. **Alter, M. and Woodruff, L. L.,** Magnetic Separation: Recovery of Salable Iron and Steel from Municipal Solid Waste, EPA/530-599, U.S. Environmental Protection Agency, March 1977.
45. **Tobert, R.,** Belt type magnets or drum magnets: which best serves resource recovery?, *Solid Waste Manage./RRJ,* 19(2), 18, 1976.

INDEX

A

B

C

D

E

F

G

H

I

K

L

M

W

Z